Integrating Human Factors Methods and Systems Thinking for Transport Analysis and Design

HUMAN FACTORS OF SIMULATION AND ASSESSMENT

Series Editors:

Dr. Michael G. Lenné

Monash University Accident Research Centre, Australia

&

Dr. Mark Young

Loughborough Design School, Loughborough University, UK

Integrating Human Factors Methods and Systems Thinking for Transport Analysis and Design

Gemma J. M. Read
Vanessa Beanland
Michael G. Lenné
Neville A. Stanton
Paul M. Salmon

CRC Press
Taylor & Francis Group
Boca Raton London New York

CRC Press is an imprint of the
Taylor & Francis Group, an **informa** business

CRC Press
Taylor & Francis Group
6000 Broken Sound Parkway NW, Suite 300
Boca Raton, FL 33487-2742

© 2017 by Taylor & Francis Group, LLC
CRC Press is an imprint of Taylor & Francis Group, an Informa business

No claim to original U.S. Government works

Printed on acid-free paper

International Standard Book Number-13: 978-1-1387-4923-8 (Paperback)
International Standard Book Number-13: 978-1-4094-6319-1 (Hardback)

Library of Congress Cataloging-in-Publication Data

Names: Read, Gemma J. M., author.
Title: Integrating human factors methods and systems thinking for transport
analysis and design / Gemma J.M. Read, Vanessa Beanland, Michael G.
Lenné, Neville A. Stanton, Paul M. Salmon.
Description: Boca Raton : Taylor & Francis, CRC Press, 2017. | Series: Human
factors of simulation and assessment | Includes bibliographical references.
Identifiers: LCCN 2017002249| ISBN 9781409463191 (hardback : alk. paper) |
ISBN 9781315589022 (ebook)
Subjects: LCSH: Highway-railroad grade crossings--Safety measures. |
Roads--Design and construction--Human factors. | Traffic signs and
signals. | Automobile drivers--Psychology. | Railroad
accidents--Prevention. | Traffic accidents--Prevention.
Classification: LCC TF263 .R43 2017 | DDC 388.3/122--dc23
LC record available at https://lccn.loc.gov/2017002249

Visit the Taylor & Francis Web site at
http://www.taylorandfrancis.com

and the CRC Press Web site at
http://www.crcpress.com

Printed and bound in Great Britain by
TJ International Ltd, Padstow, Cornwall

Dedication

For our colleagues Tom and Eric, and for all those who have been affected by crashes at rail level crossings

Contents

SECTION I   Introduction to the Research Approach

SECTION II   Rail Level Crossing Data Collection and Analysis

SECTION III   Design of New Rail Level Crossing Environments

SECTION IV   Evaluation of Design Concepts

SECTION V   Conclusions and Future Applications

Preface

In April 2006, a truck carrying a 14-tonne slab of granite was struck by a passenger train as it traversed a rail level crossing in the rural area of Trawalla, Victoria, Australia. Signage at the crossing required road users to stop and give way to trains, but there were no flashing lights or boom barriers installed. As a result of the collision, the train driver and two train passengers were killed. Just over a year later in June 2007, approximately 230 km from the site of the Trawalla collision, the driver of a loaded articulated truck was travelling at 100 km/h along a familiar rural highway near Kerang. Apparently not noticing the flashing lights or approaching train, the truck driver continued towards the crossing with the intention of driving through. Although the driver eventually saw the train and attempted evasive action, the truck struck the second carriage of the train, resulting in the deaths of 11 train passengers.

These tragic events represented a wake-up call for the Australian rail industry. Previously, crashes at rail level crossings were viewed as a road transport problem, but the emergence of significant numbers of train casualties in level crossing collisions gave rise to renewed concern. It was clear that existing rail level crossing designs were unable to ensure safety, and that human factors considerations were not well integrated into the design, assessment and operation of these legacy systems.

The design of warnings and technologies at rail level crossings has a strong historical basis. For example, the flashing red lights provided to warn of approaching trains were designed to resemble a red lantern being swung from side to side, as this is how signalmen or station masters warned the road users of that time, such as horse-drawn cart drivers, of approaching trains prior to the introduction of electric track circuits (Green 2002). Similarly, modern train horns were designed to emulate the sound of steam train whistles (Transportation Safety Board of Canada 1996), rather than purposefully designed to provide an optimal auditory warning to road users. The continuing relevance of these designs to modern transportation systems has rarely been questioned, even though the context of their use has changed considerably with increased rail and road traffic, increasing trends towards active transport modes, changes to surrounding road infrastructure and improvements to vehicle design and capabilities. Whereas road and rail systems have evolved and become far more complex, rail level crossing infrastructure and warning systems have not necessarily kept up.

With heightened concerns over safety issues at rail level crossings, it was recognised by government and industry that a collaborative approach was required. Human factors professionals and behavioural scientists employed in road authorities, rail authorities and rail operators formed a committee to coordinate human factors research initiatives, review data trends and investigation findings and provide expertise into research and development of new technologies. This committee commissioned a literature review into human factors issues at level crossings and, when the review was completed, it was evident that answers would not be found in the existing literature (Edquist et al. 2009).

Indeed, it was clear that a new approach was required. In response, the project leaders proposed *systems thinking* as a promising way forward to address this intractable problem. In particular, Cognitive Work Analysis, a systems analysis and design approach that members of the research team had used previously to redesign other complex systems, was identified as a suitable methodological framework. Accordingly, it was proposed as a key methodology for evaluating the current rail level crossing system and generating new designs to reduce the risk of collisions and subsequent fatalities and injuries. Such an approach was highly novel and innovative at the time, as it had not previously been applied in the rail level crossing context.

The research proceeded with funding from the Australian Research Council and our industry partners through a Linkage Project grant (LP100200387). A project management committee was established in 2011 to oversee the research programme, comprising representatives from the key Victorian government agencies and industry organisations. The first-of-its-kind research programme was completed in 2016, creating new insights into behaviour at rail level crossings along with new knowledge on how rail level crossing safety can be improved.

The purpose of this book is to share the approach taken over the multi-year research programme and to communicate the key findings around rail level crossing safety and behaviour. Accordingly, the book covers the data we collected, the methods we applied and how we engaged with our government and industry stakeholders, along with the key findings from each stage of the research. We also provide suggestions on how such approaches could be adopted or adapted in future research to address other transport and wider societal problems.

WHO SHOULD READ THIS BOOK?

This book is intended to be of interest to academic and industry researchers, postgraduate students and human factors practitioners who are faced with solving complex issues and problems in the transportation industries. We believe that the approaches are useful for transportation systems generally for optimising the interactions of humans and technology across the system life cycle – from design, construction and commissioning, to operation, maintenance and decommissioning.

We hope that experienced human factors researchers and practitioners will find some new methods, insights and learnings from reading the material and that researchers new to human factors and/or systems thinking will find useful guidance and advice.

Naturally, the book also outlines our findings in relation to rail level crossings. We expect that these will be of interest to those working in this area, and we hope that all readers will find the research findings as interesting and thought-provoking as we have while undertaking this work.

WHY SHOULD YOU READ THIS BOOK?

There are a number of excellent books available that provide guidance on the use of human factors and systems thinking methods (e.g. Crandall, Klein and Hoffman 2006, Naikar 2013, Stanton et al. 2013, Vicente 1999). Further, it is widely acknowledged that multi-method approaches are needed to understand and address complex problems such as accidents in transportation systems. Despite this, there is little guidance on how to select appropriate methods and to integrate them within a single research project that spans analysis, design and evaluation. This book intends to address this gap and to provide you with tools and advice for taking a similar approach to solve other problems in transportation and beyond.

HOW TO READ THIS BOOK

We expect that some readers will be highly familiar with the methods and approaches discussed, whereas for others there will be much new information. We have tried to achieve a balance in the level of detail provided and, where possible, refer the novice reader to other texts that they may find useful for further guidance and exemplars.

We use examples from our work in rail level crossing analysis, design and evaluation to illustrate the approach throughout the book. However, you will find other examples discussed throughout the book and general principles highlighted that can be applied to broader transport issues as well as domains outside of transportation.

This book is divided into five main sections:

I. *Introduction to the Research Approach*
   • Chapters 1 and 2 provide an introduction to the key human factors and systems thinking philosophies, theories and methods adopted throughout the research programme. Chapter 3 then outlines the integrated framework of methods we applied.
II. *Rail Level Crossing Data Collection and Analysis*
   • Chapter 4 describes the data collection activities we undertook to understand road user behaviour at rail level crossings. The findings reported in this chapter draw on traditional human factors analysis methods.
   • Chapter 5 describes how the data were used to develop systems-based models that describe the functioning of rail level crossings.
III. *Design of New Rail Level Crossing Environments*
   • Chapters 6 through 8 describe the process undertaken to generate novel designs for rail level crossings, to conduct an initial desktop evaluation using the systems thinking models described in Chapter 5 and to subsequently refine the designs.

IV. *Evaluation of Design Concepts*
- Chapter 9 describes a suite of driving simulator studies that were used to measure driver responses to the new designs.
- Chapter 10 presents the findings of a survey undertaken to elicit feedback from all road user types (i.e. car drivers, heavy vehicle drivers, motorcyclists, cyclists and pedestrians) on the proposed designs.

V. *Conclusions and Future Applications*
- Chapter 11 outlines the recommendations arising from the research, identifies future research directions relevant to rail level crossing safety and reflects upon the extent to which the research programme met its overall aims.
- Chapter 12 discusses potential future applications of the research approach to other transport and non-transport domains.

We hope that this book might inspire new applications of human factors and systems thinking that continue to extend the methods and approaches adopted as well as to provide practical recommendations that can help to address real-world problems.

Acknowledgements

We acknowledge the many individuals and organisations that contributed to this programme of research. The research was funded through an Australian Research Council Linkage Grant (LP100200387) provided to the University of Sunshine Coast, Monash University, and the University of Southampton, in partnership with the following organisations: the Victorian Rail Track Corporation (VicTrack), Transport Safety Victoria, Public Transport Victoria, Transport Accident Commission, Roads Corporation (VicRoads) and V/Line Passenger Pty Ltd. We also appreciate the support provided by Metro Trains Melbourne and the Office of the Chief Investigator, Transport Safety.

In particular, we acknowledge the support and input of the project management committee members Todd Bentley, Samantha Buckis, Ben Cook, Ian Davidson, Sarah-Louise Donovan, Nicholas Duck, Elizabeth Grey, Kelly Imberger, Dean Matthews, Chris McKeown, Peter Nelson-Furnell, Darren Quinlivan, Terry Spicer and Ash Twomey. It should be noted that the views and opinions expressed in this book are those of the authors and do not necessarily represent the views of the partner organisations or their representatives.

We express our great appreciation for the contributions of staff and research students at Monash University and the University of the Sunshine Coast who worked on this research programme. We thank Christine Mulvihill, Nicholas Stevens, Eryn Grant, Amanda Clacy, Ashleigh Filtness, Kristie Young, Miles Thomas, Tony Carden, Miranda Cornelissen, Casey Rampollard, Michelle van Mulken, Nebojsa Tomasevic, Christian Ancora, Ash Verdoon, Kerri Salmon, Jessica Edquist, Russell Boag and Georgia Lawrasia. We also thank Guy Walker from Heriot-Watt University for his extensive contributions to the research programme.

Importantly, we take this opportunity to acknowledge the significant contributions of Professor Thomas Triggs for his valuable input to the initial research design and conception for this project and Dr. Eric Wigglesworth AM for his input into research to support the grant application. Both Tom and Eric have left a legacy in rail level crossing human factors research, and we are honoured to have had the opportunity to work with them in this important area.

We thank the VicRoads Northern Region Office for making their office space available as a base to conduct our rural on-road study and the Air Operations division of the Defence Science and Technology Group, Department of Defence, for the loan of eye tracking equipment to support our on-road studies. Finally, we thank our research participants who generously shared their time, expertise and ideas with us to support this important research.

Professor Paul Salmon's contribution to this book was funded through his Australian Research Council Future Fellowship (FT140100681). Dr. Vanessa Beanland's contribution to this book was funded through her Australian Research Council Discovery Early Career Researcher Award (DE150100083).

About the Authors

Gemma J. M. Read, PhD, is a research fellow at the Centre for Human Factors and Sociotechnical Systems at the University of the Sunshine Coast, Queensland, Australia. She completed her PhD in human factors at the Monash University Accident Research Centre, Victoria, Australia, in 2015, focussed on the development of a design toolkit for use with Cognitive Work Analysis. Prior to this, she completed a postgraduate diploma in psychology and undergraduate degrees in behavioural science and law. Gemma has more than 10 years' experience applying human factors methods in transportation safety, predominantly within the area of transport safety regulation.

Vanessa Beanland, PhD, is a research fellow in the Centre for Human Factors and Sociotechnical Systems at the University of the Sunshine Coast, Queensland, Australia. She completed her PhD in experimental psychology in 2011 and previously worked as a lecturer at the Australian National University, Canberra, Australia, and a research fellow at the Monash University Accident Research Centre, Victoria, Australia. Vanessa currently holds an Australian Research Council Discovery Early Career Research Award in the area of driver behaviour and has more than 8 years' research experience encompassing applied psychology and human factors in transport.

Michael G. Lenné, PhD, is an adjunct professor (research) at the Monash University Accident Research Centre, Victoria, Australia. He obtained a PhD in human factors psychology in 1998 and has since worked in a number of research roles in university and government settings, all with a focus on improving transport safety. For 7 years, he had led the human factors research team at the Monash University Accident Research Centre, and in 2014, he became a professor in human factors. In late 2014, his unique expertise in human factors and research translation were recognised through the creation of the role of chief scientific officer, human Factors at Seeing Machines Ltd, Canberra, Australia. Mike has dedicated his research career to measuring human operator performance in transport settings and in developing and evaluating measures to improve safety.

Neville A. Stanton, PhD, DSc, is a chartered psychologist, chartered ergonomist and chartered engineer. He holds the chair in human factors engineering in the Faculty of Engineering and the Environment at the University of Southampton, Southampton. He possesses degrees in occupational psychology, applied psychology and human factors engineering. His research interests include modelling, predicting, analysing and evaluating human performance in systems as well as designing the interfaces and interaction between humans and technology. Neville has worked on the design of automobiles, aircraft, ships and control rooms over the past 30 years, on a variety of automation projects. He has published 35 books and over 270 journal papers on ergonomics and human factors. His work has been recognised through numerous awards, including the Hodgson Prize, awarded to him and his colleagues in 2006 by the Royal Aeronautical Society, London, for research on design-induced, flight-deck error. He has also received numerous awards from the Chartered Institute of Ergonomics and Human Factors in the United Kingdom and in 2014 the University of Southampton, Southampton, awarded him a Doctor of Science for his sustained contribution to the development and validation of human factors methods.

Paul M. Salmon, PhD, is a professor in human factors, and is the creator and direc-tor of the Centre for Human Factors and Sociotechnical Systems at the University of the Sunshine Coast, Queensland, Australia. He currently holds an Australian Research Council Future Fellowship in the area of transportation safety and leads major research programmes in the areas of road and rail safety, cybersecurity, and led outdoor recreation. His current research interests include accident prediction and analysis, distributed cognition, systems thinking in transportation safety and health care and human factors in elite sports and cybersecurity. Paul has coau-thored 11 books, more than 150 journal articles and numerous conference articles and book chapters. He has received various accolades for his contributions to research and practice, including the UK Ergonomics Society's President's Medal, the Royal Aeronautical Society's Hodgson Prize for best research and paper and the University of the Sunshine Coast's Vice Chancellor and President's Medal for Research Excellence. In 2016, Paul was awarded the Australian Human Factors and Ergonomics Society Cumming Memorial Medal for his research contribution.

Section I

Introduction to the Research Approach

1 Systems Thinking in Transport Analysis and Design

1.1 INTRODUCTION

Surface transportation systems, such as road and rail, continue to evolve at a rapid rate of change. This evolution is driven by the ubiquitous introduction of new and advanced technologies, significant increases in the intensity of operations and the presence of diverse types of end users. Although many of the changes are underpinned by a desire to prevent fatalities and injuries, and system changes to date have produced important gains in terms of reduced crashes and trauma, in many jurisdictions we are seeing these numbers plateau. Indeed, the year 2016 in Australia saw an increase in the road toll in several states (BITRE 2016), with similar trends in other international jurisdictions such as the United States (National Safety Council 2016) and the United Kingdom (Department for Transport 2016).

One longstanding surface transportation issue is collisions at rail level crossings. Such collisions represent a persistent source of road trauma, with collisions between road vehicles and trains accounting for approximately 45% of rail fatalities in Australia (ONRSR 2015). To ensure that future transport systems are as safe as expected by the community, a new approach is required to facilitate the removal of longstanding issues and the management of emergent problems.

In response to this, and in line with trends in other safety-critical domains, road and rail safety is beginning to see a shift in the way that safety problems are analysed and addressed. Central to this shift is the argument that existing deterministic approaches do not fully consider the inherent complexity in transportation systems, nor the full range of factors shaping behaviour (Cornelissen et al. 2015, Larsson et al. 2010, McClure et al. 2015, Salmon and Lenné 2015, Salmon et al. 2012b). Traditionally, transport safety practitioners adopted a deterministic, reductionist approach to safety issues. This involved decomposing transport systems into their component parts (e.g. drivers, vehicles, warning devices), examining the parts in isolation (e.g. driver errors, vehicle safety systems, warnings), attempting to improve the performance of parts (e.g. preventing driver errors, making vehicles safer, making a warning more conspicuous) and then reintroducing them back into the system. Although this approach produced some successful interventions (e.g. seatbelts, airbags), it gave rise to a fascination with 'human error' along with a fixation on the human operator as the primary cause of transportation crashes (e.g. Reason et al. 1990). For over four decades, headlines describing the high proportion of transportation crashes caused by driver error have been the norm. As a corollary, many safety interventions in this time focussed predominantly on drivers and aimed to improve their behaviour through education, training, enforcement or the prohibition of undesirable behaviours. Although parts of the transport systems were improved, little consideration was given to how these parts interact with one another, or how the overall transport system functions.

1.1.1 Systems Thinking and Rail Level Crossings

In the area of rail level crossings, researchers have highlighted an inadequate understanding of behaviour at rail level crossings and, more specifically, a lack of understanding of the interactions between road users (including pedestrians) and the crossing infrastructure that give rise to unsafe behaviours (Edquist et al. 2009). It has been argued that interactions such as these are best understood using a systems thinking approach (Salmon and Lenné 2015; see Box 1.1). The reductionist approach has been criticised not only because it examines component parts in isolation, but also because it artificially separates the rail and road infrastructures, with each managed by different parties. It is contended that the solution to the rail level crossing problem lies in embracing systems thinking approaches to firstly better understand the problem, and then apply this understanding to design new solutions that optimise system functioning.

Based on this, the aim of the programme of research described in this book was to apply a systems approach to the problem of rail level crossings. This chapter will set the scene for these applications by outlining the systems approach and demonstrating it in the rail level crossing safety context.

BOX 1.1   SYSTEMS THINKING AND SAFETY

A systems thinking approach to safety involves taking the overall system as the unit of analysis, looking beyond the individual and considering the interactions between humans and between humans and technology within a system. This view also considers factors relating to the broader organisational, social or political system in which processes or operations take place. Taking this perspective, safety emerges not from the decisions or actions of an individual, but from interactions between humans and technology across the wider system. In the rail level crossing context, this means that decisions and actions made at government, regulatory and rail operating company levels all play a role in collisions. This calls for a more comprehensive approach to analysis and design that goes beyond road users and the physical rail level crossing environment.

1.2 UNDERSTANDING THE SYSTEMS THINKING APPROACH

The term 'systems thinking' is used throughout this book to describe a philosophy currently prevalent within the discipline of human factors, which aims to understand and improve the performance and safety in complex sociotechnical systems. It is most prominent in the area of accident analysis and prevention whereby, after first emerging in the early twentieth century (e.g. Heinrich 1931), it is now characterised by a series of accident causation models and analysis methods (e.g. Leveson 2004, Perrow 1999, Rasmussen 1997, Reason 1997, Svedung and Rasmussen 2002). Contemporary models are underpinned by the notion that safety and accidents are emergent properties arising from non-linear interactions between multiple components across complex sociotechnical systems (e.g. Leveson 2004).

Perhaps facilitated by the ubiquity of systems thinking within safety science circles, the past decade has also seen increasing applications of sociotechnical systems theory for system design and analysis (Eason 2014). Underpinned by systems thinking, sociotechnical systems theory is a work design approach concerned with both the performance of the work system and the experience and well-being of workers (Clegg 2000). A key tenet of sociotechnical systems theory is that systems require adaptive capacity; one of the primary means to achieve this is through joint optimisation of human and technical elements across the system of interest. Jointly optimising systems requires adoption of a systems thinking approach when examining behaviour and when identifying the ways of improving behaviour. This characteristic ensures that sociotechnical systems theory and methods are highly compatible with the systems thinking approach.

In recent times, the potential utility of both systems thinking and sociotechnical systems theory for transportation system analysis and design has been recognised. There is now growing consensus that increased reductions in trauma may be facilitated by integrating and applying systems thinking and sociotechnical systems theory approaches in this context. Specifically, this includes applying models of system safety and accident causation (e.g. Rasmussen 1997), STS design principles (e.g. Clegg 2000, Davis et al. 2014) and systems analysis and design methodologies (e.g. Checkland 1981, Hollnagel 2012, Leveson 2004, Stanton et al. 2013b, Sterman 2000, Svedung and Rasmussen 2002, Vicente 1999). The overriding philosophy underpinning these approaches is that it is the entire system that needs to be optimised (i.e. the overall rail level crossing system), not just the individual components acting within it (e.g. road users).

Although the systems approach quickly gained widespread acceptance in many safety critical domains, it has not gained traction in road and rail safety until recently, when various researchers began to discuss its merits (Larsson et al. 2010, Read et al. 2013, Salmon and Lenné 2009, 2015, Salmon et al. 2012b) and conduct exploratory applications (Cornelissen et al. 2013, Goh and Love 2012, Newnam and Goode 2015, Salmon et al. 2013b, Young and Salmon 2015). Put simply, the arguments have centred on the notion that road crashes have systemic causes over and above those related to driver behaviour. A key tenet of the systems thinking approach is that driver errors in fact represent the consequences of system-wide issues, rather

than the primary cause of crashes. Therefore, holistic interventions should be introduced across all aspects of transport systems, as opposed to traditional approaches
of fixing components (i.e. drivers, vehicles or the road environment) in isolation.
Examples of holistic interventions include modifications to policy, certification,
regulatory frameworks and standards and guidelines. These interventions should be
carefully designed to optimise system functioning, rather than representing piecemeal responses to address isolated flaws identified following an accident.

### 1.2.1	Rasmussen's Framework

Rasmussen's risk management framework (Rasmussen 1997; see Figure 1.1) is one
popular systems thinking model that is beginning to be applied in road and rail
safety settings (e.g. Newnam and Goode 2015, Salmon et al. 2013c, Scott-Parker
et al. 2015, Young and Salmon 2015). The framework argues that systems comprise

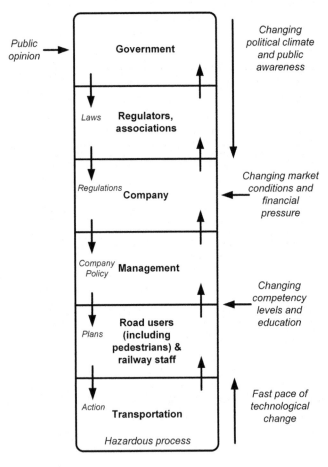

FIGURE 1.1 Rasmussen's risk management framework adapted for rail level crossings.
(Adapted from Rasmussen 1997.)

various hierarchical levels (e.g. government, regulators, company, company management, staff and work/activity), each of which contains actors (individuals, organisations or technologies) who are co-responsible for production and safety. Decisions and actions occurring at all levels interact to shape system performance, meaning both safety and accidents are influenced by the decisions of all actors, not just front-line workers and operators. Further, the framework argues that accidents are caused by multiple contributing factors, not just one bad decision or action. A key implication is that it is not possible to truly understand safety and performance by decomposing the system and examining its components alone; rather, it is the interaction between components that are of interest. Further, the more components and interactions studied together, the closer one can get to understanding system performance and the factors influencing it.

As described earlier, the prevalent approach in road and rail safety has been reductionist, focussing mainly on components such as individual road users attempting to improve their behaviour (Cornelissen et al. 2013, Larsson et al. 2010, Read et al. 2013, Salmon et al. 2012b). Although a focus on road users is important, systems thinking argues that the road user is merely one part of a rich interconnected network of human and technical components, and therefore, road user behaviour must be studied and optimised in the broader context of the interactions with other system components.

1.2.2 SOCIOTECHNICAL SYSTEMS THEORY

Sociotechnical systems theory emerged in the 1950s from a programme of research undertaken at the Tavistock Institute, London, UK, which focussed on the disruptive impacts of new technologies on human work (Eason 2008, Trist and Bamforth 1951). Primarily a work design theory, sociotechnical systems theory is heavily underpinned by systems theory and contains principles related to participative democracy and humanistic values. This engenders a focus on both the performance of the work system and the experience and well-being of the people performing the work (Clegg 2000). A key contribution of sociotechnical systems theory is the provision of various values and principles to support the design of sociotechnical systems that align with open systems principles (e.g. Cherns 1976, Clegg 2000, Davis 1982, Walker et al. 2009b).

Being underpinned by systems theory, sociotechnical systems theory shares the notion that it comprises both social and technical elements co-engaged in the pursuit of shared goals. The interaction of these social and technical aspects creates emergent properties and the conditions for either successful or unsuccessful system performance (Walker et al. 2009b). Accordingly, joint optimisation – as opposed to the optimisation of solely the social or technical aspects – is required for safe and efficient system performance (Badham et al. 2006).

Based on many applications in work design, there is strong evidence that applying sociotechnical systems theory values and principles can have a series of benefits. For example, a meta-analysis of 134 studies involving applications of sociotechnical systems theory found that almost 90% reported improvements in safety and productivity and >90% reported improvements in workers' attitudes and quality of outputs (Pasmore et al. 1982). Until now, the approach has been applied overwhelmingly to the introduction of new technologies (such as computer systems) within organisations (Davis et al. 2014). Proponents of the sociotechnical approach have called for its expansion to the entire work system, including the design of physical working environments (Davis et al. 2011), as well as to broader societal issues that span multiple organisations such as security, sustainability, health-care provision and urban planning (Davis et al. 2014). Prior to the work described in this book, to our knowledge sociotechnical systems theory had not been applied to the design of surface transportation systems.

1.3 HOW DOES STS AND THE SYSTEMS THINKING APPROACH APPLY TO RAIL LEVEL CROSSING COLLISIONS?

The research programme described in this book involved applying systems thinking and sociotechnical systems theory and methods, in an integrated manner, to rail level crossing system analysis and design. When considered together in this manner, the systems thinking and sociotechnical systems theory approaches provide three key assertions that demand a different approach to rail level crossing safety:

1. All sociotechnical system comprise a series of hierarchical levels containing multiple actors and organisations that are co-responsible for safety.
2. Accidents are systems phenomena involving multiple interacting contributory factors, and these factors emerge and reside across the different hierarchical levels.
3. The attainment of safe and efficient systems requires joint optimisation of human and technical elements.

The application of these assertions to the rail level crossing system is discussed in Sections 1.3.1 and 1.3.2.

1.3.1 THE RAIL LEVEL CROSSING SYSTEM

According to Rasmussen's (1997) risk management framework, the rail level crossing system can be viewed as comprising a series of hierarchical levels, each containing multiple actors and organisations that ultimately work together to create rail level crossing safety. Figure 1.2 presents an 'ActorMap' of the rail level crossing system in Victoria, Australia. Based on Rasmussen's risk management framework, the ActorMap details each level of the system and identifies the actors who share the responsibility for rail level crossing safety. An important implication here is that any rail level crossing collision is effectively created by a network of interacting decisions and actions made by actors across all levels of the ActorMap.

1.3.2 RASMUSSEN'S ACCIDENT CAUSATION TENETS

Rasmussen's framework incorporates a series of tenets regarding accident causation that provide a valuable framework for studying safety issues in various domains

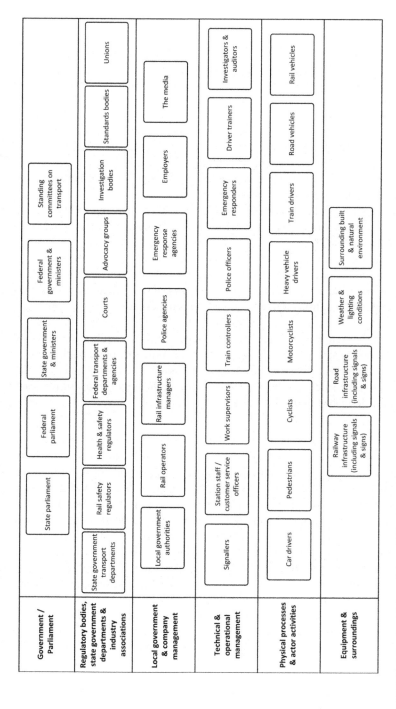

FIGURE 1.2 ActorMap of the rail level crossing system in Victoria, Australia.

(Vicente and Christoffersen 2006). These tenets can be adapted to fit the rail level crossing context:

1. Collisions at rail level crossings are emergent properties impacted by the decisions and actions of all actors across road and rail transport systems, not just road users alone.
2. Threats to rail level crossing safety are caused by multiple contributing factors, not just a single poor decision or action made by an individual road user or train driver.
3. Threats to rail level crossing safety can result from poor communication across levels of the system (i.e. a lack of 'vertical integration'), not just from deficiencies occurring at one level alone.
4. Lack of vertical integration is caused, in part, by lack of feedback across levels of the road and rail transport system.
5. Activities associated with maintaining rail level crossing safety are not static; they migrate over time and under the influence of various pressures such as financial, production and psychological pressures.
6. Migration occurs at multiple levels of road and rail transport systems.
7. Migration of activities causes system defences to degrade and erode gradually over time, not all at once. Rail level crossing collisions are caused by a combination of this migration and a triggering event/s.

Along with the ActorMap in Figure 1.2, the tenets have several key implications for rail level crossing safety research and practice. It is precisely these implications that, in our view, demand a paradigm shift in how we attempt to understand and enhance rail level crossing safety. In this sense, they formed the foundations for the research programme described in this book.

1. When attempting to understand and prevent rail level crossing collisions, research and practice should focus on the decisions and actions made by all actors within road and rail transport systems, not just those made by road users (e.g. road vehicle drivers, pedestrians). Even when factors such as distraction, speeding or impairment were involved in a rail level crossing collision, there are still underlying behaviours and interactions across the system that enabled the collision to occur. The key to optimising rail level crossing safety lies in understanding which factors interact to create rail level crossing collisions.
2. Interventions should focus on optimising human and technical elements across all levels of road and rail transport systems, not just end users. Historically, level crossing safety has focussed on road vehicle drivers and/or pedestrian behaviour, and has primarily included interventions that improve behaviour (Read et al. 2013). Although some interventions have proven successful, Rasmussen's framework suggests that the decisions and actions of others within the system must also be considered to maximise the potential benefits. Importantly, the decisions and actions of those at the

higher levels of the road and rail transport system potentially have a greater influence on overall safety. Such actors include policy makers, regulators, designers and engineers, road safety authorities, train operators and track owners, to name a few. Important requirements then are to understand what role each actor plays in rail level crossing safety and how the actors interact during rail level crossing system design and operation.

3. The extent of vertical integration present in rail level crossing systems requires investigation. Interactions across different levels of road and rail transport systems have received little attention to date, yet they could conceivably shed light on rail level crossing collisions. Research and practice should therefore aim to understand and enhance communication and feedback across road and rail transport systems.

4. The pressures and constraints that influence rail level crossing functioning, as well as behaviour at different levels of road and rail transport systems, need to be identified. Although financial, production and psychological pressures undoubtedly play a role, less is known about other related pressures, including political, social and organisational constraints. Without clarifying these factors and their impact, it is impossible to prevent migration of behaviour and safety towards that which is unacceptable.

1.4 SYSTEMS THINKING APPLIED: THE CRASH AT KERANG

In June 2007, a loaded semi-trailer truck struck a passenger train on a rail level crossing near the town of Kerang in northern Victoria, Australia, killing 11 train passengers and injuring a further 14 passengers and the truck driver. We conducted an analysis of this event based on the outcomes of the independent investigation by the Office of the Chief Investigator, Victoria (OCI) and the transcript of criminal proceedings following the prosecution of the truck driver (who was subsequently acquitted of all charges by a jury). Our full analysis is presented in the work of Salmon et al. (2013c); however, a summary is presented here.

At approximately 10:30 a.m. on Tuesday, 5 June 2007, an articulated truck comprising a prime mover and a trailer departed its depot in Wangaratta, Victoria, embarking on a weekly 820-km freight run to Adelaide, South Australia. The driver had more than 25 years' experience of truck and van driving, and a good driving record with no infringements (OCI 2007). He was also familiar with the route, having driven it around once a week for approximately 7 years. Finally, the truck driver had just returned to work that day following 4 weeks leave (OCI 2007).

At approximately 1:00 p.m., a regional passenger train departed Swan Hill station in Victoria with five planned stops *en route* to Melbourne. The train comprised a locomotive hauling a three-car passenger set. After travelling through 26 rail level crossings, the train approached crossing Y2943 on the Murray Valley Highway. Approximately 25 s before it reached the crossing, the train passed over a track circuit, activating the crossing's flashing lights and warning bells. Following this, the train driver activated the air horn as he approached the whistle board.

Travelling at around 100 km/h, the truck passed the 'RAIL' and 'X' road markings (approximately 267 and 253 m from the crossing, respectively), a rail level crossing warning sign (approximately 260 m from the crossing) and continued to approach the crossing without slowing. The truck driver later stated that, after noticing the warning sign, he checked the flashing light assembly, but did not detect the lights flashing (R v Scholl 2009).

Approximately 140 m before the crossing, the train driver noticed that the truck, which was now approximately 70–100 m away, was approaching the crossing at speed. In response, he sounded the train horn for several seconds. After noticing a stationary vehicle waiting on the other side of the crossing, the truck driver finally became aware of the train, applied the brakes, and attempted to take an evasive action by steering the truck into a gully to the left of the train tracks. At approximately 1:34 p.m., the truck struck the second passenger car of the train. The collision led to the train derailing and caused substantial damage to the truck.

As noted, several warnings were in place to prevent a vehicle–train collision at the rail level crossing (see Figure 1.3). These included road-based warnings (e.g. road markings and road signage), rail level crossing-based warnings (e.g. flashing lights and warning bells) and train-based warnings (e.g. train horn).

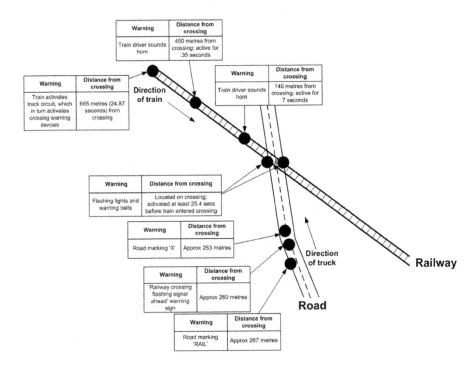

FIGURE 1.3 Kerang rail level crossing warnings. (Reprinted from *Accident Analysis and Prevention*, Vol. 50, Salmon, P. M., G. J. M. Read, N. A. Stanton, and M. G. Lenné, The crash at Kerang: Investigating systemic and psychological factors leading to unintentional non-compliance at rail level crossings, Pages 1278–1288, Copyright (2013), with permission from Elsevier.)

The flashing lights and warning bells failed to alert the truck driver about the presence of the train. The first sounding of the train horn also failed to alert the driver. Of the other warnings, the court proceedings indicate that the truck driver saw the first crossing warning sign and proceeded to look at the crossing, so this was deemed to be effective in fulfilling its role as a passive warning device. However, the sign directed the truck driver's attention towards the flashing lights assembly, rather than prompting him to scan the tracks, so it was only partially effective. It is unclear whether the road markings 'RAIL' and 'X' were noticed by the truck driver and whether they initiated further scanning of the crossing. Finally, the second sounding of the train horn, 120 m from the crossing, also failed to alert the truck driver about the presence of the train. Notably, the cue that finally alerted the driver to the train's approach – namely the other vehicle stopped at the crossing – was not an inherent design feature of the rail level crossing system, but rather was an emergent property from interactions within the system.

1.4.1 THE INDIVIDUAL PERSPECTIVE

From an individual component or reductionist perspective, it is relatively straightforward to explain the Kerang incident. This viewpoint tells us that the crash was caused by the truck driver's failure to see the flashing lights or approaching train until it was too late to successfully execute evasive manoeuvres. In turn, this failure has been explained via the psychological phenomena of a so-called looked-but-failed-to-see (LBFTS) error. Salmon et al. (2013c) describe how the driver's extensive experience of the crossing with no train present led to the development of schemata, or mental models of the world as we expect it to behave, which effectively enabled the driver to look at the warnings but not perceive that they were active (Salmon et al. 2013c). Although this explanation is appropriate, it does not consider other important factors such as why the crossing did not have full boom gate protection (to prevent an LBFTS error) or the factors that exacerbated the LBFTS error. Indeed, psychological studies investigating LBFTS errors have found that these errors are heavily situation dependent, with some situations yielding LBFTS errors in 100% of individuals (Most et al. 2001). Additionally, attempts to find individual differences that predict LBFTS errors – the notion that some people are particularly susceptible to error – have generally failed (Beanland and Chan 2016, Bredemeier and Simons 2012, Kreitz et al. 2015, 2016, Wright et al. 2013), indicating that system design is highly implicated in these errors.

1.4.2 A SYSTEMS PERSPECTIVE ON KERANG

As described previously, the systems thinking viewpoint goes beyond the individual perspective to identify various factors that interacted to cause the LBFTS error and the accident overall. These factors are represented in Figure 1.4, which presents a summary of our AcciMap of the incident. Based on the hierarchical levels outlined by Rasmussen's (1997) risk management framework (Figure 1.1), the AcciMap depicts the interconnected contributory factors across the road and rail systems.

The AcciMap demonstrates the systemic nature of the contributory factors involved in the tragic crash at Kerang. At the equipment and the surroundings level,

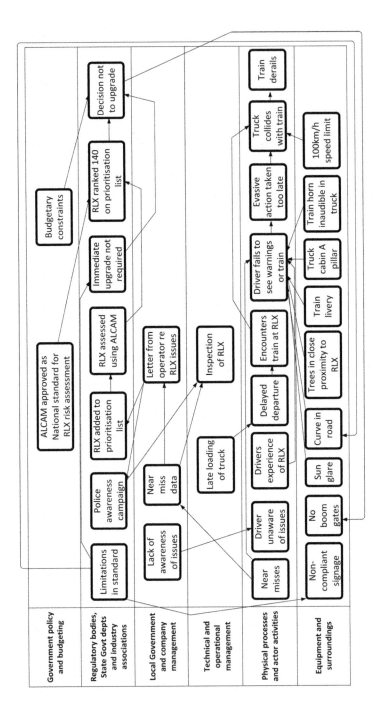

FIGURE 1.4 Kerang AcciMap. ALCAM, Australian Level Crossing Assessment Model; RLX, rail level crossing. (Adapted from Salmon, P. M. et al., *Accid. Anal. Prev.*, 50, 1278–1288, 2013.)

there are multiple factors that likely contributed to the driver's failure to notice the activated rail level crossing and the approaching train. The crossing had flashing lights and various warnings (e.g. signage, road markings) but was not fitted with boom gates. Previous research has shown that crossings with boom gates achieve the best safety performance (Saccomanno et al. 2007), primarily because they provide a more conspicuous visual cue to the driver, as well as forming a physical barrier. Weather conditions at the time of the incident are also important; the OCI report states that the sun was directly in front of the truck and a test run in similar weather conditions reported considerable sun glare from the road surface (OCI 2007). The OCI also reported that the contrast between the train and its background is likely to have been reduced because the truck-facing side of the train was shadowed. Trees near the crossing may have obscured the truck driver's vision of the approaching train, whereas the A-pillar of the truck also momentarily obscured a stationary vehicle located on the opposite side of the crossing. Although the train driver sounded the train horn twice on approach to the crossing, the OCI stated that it is unlikely that the first horn sounding would have been audible to the truck driver (OCI 2007). Finally, the road speed limit was 100 km/h; a lower travel speed would have provided more time for evasive action.

The truck driver's experience of the route also played a significant role in the incident. Critically, despite driving the same route more than 300 times previously, he had not previously experienced a train at the Kerang crossing (R v Scholl 2009). On the day of the incident, the driver was delayed in departing the depot due to freight loading issues, which in turn led to the truck encountering the train at the crossing. Influenced by a strong expectancy for the crossing to be inactive with no train approaching, along with the environmental, meteorological and vehicle factors noted previously, the driver failed to notice the approaching train or the activated warnings.

As we move further up the system, we begin to see important contributory factors that occurred in the weeks and months prior to the incident. An infrastructure manager conducted an inspection of the crossing in response to a series of near miss incidents, and a letter was sent from the train operator to the track manager expressing concern over road user behaviour at the crossing. Notably, the truck driver's employer was not made aware of the inspection or the near miss incidents.

Higher up in the system, the contributory factors relate to the absence of boom gates at the Kerang crossing. In response to the near-miss incidents, the crossing was added to the state government's crossing prioritisation list (OCI 2007), which meant that it would be under consideration for upgrade to full boom gate controls. Accordingly, approximately 8 months prior to the incident, the crossing was assessed using the Australian Level Crossing Assessment Model (ALCAM). At the time of the incident, the ALCAM risk assessment tool placed little emphasis on the role of human factors in behaviour at rail level crossings, and viewed risk primarily as a function of road and rail traffic levels. Accordingly, the Kerang crossing was assigned a risk score that led to it being ranked 140 out of 143 crossings on the prioritisation list (OCI 2007), which meant it would not be upgraded for some time.

This slow upgrade of crossings on the prioritisation list was a function of budgetary constraints: due to the very high costs associated with upgrading rural crossings, only approximately 20 crossings could be upgraded each year.

1.5 SUMMARY

The aim of this chapter was to provide an overview of systems thinking and socio-technical systems theory philosophies, particularly as they relate to the rail level crossing context. In doing so, the chapter has outlined the rich explanatory power of the systems thinking approach, along with its holistic approach to design and system optimisation.

The Kerang case study demonstrates how, even when driver behaviour is directly linked to a rail level crossing collision, adopting a systems thinking approach eluci-dates information on the contribution of factors across the overall road and rail system, and provides clues as to what system reforms are required to prevent a similar incident moving forward. In this case, although modifications to the crossing environment would be useful, the need for wider system reform around risk assessment, crossing prioritisation for upgrade and indeed the cost of crossing upgrades is emphasised. Notably, a short-term implication is that lower cost rail level crossing treatments are required that perform equally well as more costly upgrades to boom gates.

Taking a systems thinking approach to rail level crossing analysis and design raises a number of pertinent questions. Regarding the functioning of rail level cross-ing systems, there are knowledge gaps relating to the nature of the factors across road and rail level crossing systems that influence end user behaviour. Regarding the design of rail level crossing systems, it is unclear how human and technical ele-ments of the system can be jointly optimised. We contend that the answers to these questions lie at the core of rail level crossing safety. Integrating systems thinking and sociotechnical systems theory therefore provides a powerful framework for under-standing and optimising rail level crossing system behaviour. The programme of research described in the following chapters aimed to do exactly that by using meth-odologies from systems thinking and sociotechnical systems theory to:

1. More fully understand rail level crossing system behaviour and the fac-tors influencing road user behaviour (i.e. car drivers, heavy vehicle drivers, motorcyclists, cyclists and pedestrians).
2. Develop and test new STS-based rail level crossing design concepts intended to better prevent collisions and more fully support users in negotiating rail level crossings safely.

Although the explanatory power of the systems thinking approach is demonstrated through the Kerang accident analysis, it is important to note that systems think-ing and sociotechnical systems theory approaches are not limited only to accident analysis. There exists a range of methodologies that are suited to both the analysis of rail level crossing system performance and the design. Example methodologies underpinned by systems thinking and/or sociotechnical systems theory include the Cognitive Work Analysis (CWA) framework (Vicente 1999), Hierarchical Task Analysis (HTA; Stanton 2006), the Event Analysis of Systemic Teamwork (Stanton et al. 2013b) and Soft Systems Methodology (Checkland 1981). The programme of research described in this book utilised the CWA framework and HTA. An overview of these methods is provided in Chapter 2 and a description of how they were used in combination with other human factors and design methods is provided in Chapter 3.

2 An Overview of Key Human Factors Approaches and Methods

2.1 INTRODUCTION

The aim of this chapter is to introduce the key human factors and systems thinking methods that were employed throughout the research programme described in this book. Throughout the chapter, Boxes 2.1 through 2.12 provide examples of how the methods were used in the research. Chapter 3 then describes how these methods were integrated within the overall research programme.

As introduced in Chapter 1, a defining characteristic of this research programme was that it adopted a systems approach to rail level crossings, underpinned by the principles of sociotechnical systems theory. This necessitated adopting a multitude of research methods, both to capture performance at all levels of the system hierarchy and to address different stages of the system life cycle, ranging from analysing the existing system functioning through to redesigning the system. The various methods we adopted for the research programme can be grouped under four broad categories:

1. *Data collection methods for understanding human behaviour and performance*: These methods focus on collecting data about human performance within an existing system and include observation, verbal protocol analysis, cognitive task analysis interviews, subjective workload measurements, usability and preference measures, eye tracking and vehicle parameters. Such measures can be obtained from naturalistic or semi-naturalistic studies of behaviour, including on-road instrumented vehicle studies (see Chapter 4) or laboratory-based studies such as those undertaken in driving simulators (see Chapter 9).

2. *Data collection methods for understanding performance of the overall system*: These methods, which include document review and interviews with subject-matter experts, involve collecting data to understand other aspects of the system of interest, beyond users and their interactions with technology and the immediate environment.

3. *Systems-focused analysis methods*: The next set of methods described are analysis methods, which assist to structure raw data to gain insights into system functioning. These include network analysis, Hierarchical Task

Analysis (HTA), the Systematic Human Error Reduction and Prediction Approach (SHERPA) and Cognitive Work Analysis (CWA).

4. *Human factors design methods*: Finally, we introduce some human factors methods and tools that can assist to take a systems thinking and sociotechnical systems theory approach to design.

These methods are summarised in Figure 2.1, showing the levels of the system that are addressed by each, based on Rasmussen's (1997) risk management framework.

We hope that the discussion of these methods will provide valuable context for understanding the research presented throughout this book, as well as useful guidance for those wanting to use similar approaches in their own work. Readers who wish to use the methods are directed to the appendix, where more detailed procedural guidance is provided.

2.2 DATA COLLECTION METHODS FOR UNDERSTANDING HUMAN PERFORMANCE

2.2.1 OBSERVATION

Observational studies encompass a range of research methods used to collect data about tasks, behaviour, and environmental or contextual influences on behaviour. A fundamental attribute of observational methods is that naturalistic behaviour is captured in a usually non-intrusive, and sometimes covert, manner. Observation can be a valuable way to capture data without intervening in the task and provides an excellent opportunity to gain familiarisation with the system as it operates in the real world. Researchers can collect various types of observational data, including tasks, task sequences, task performance and verbal and physical interactions between system components (both human-human and human-technology interactions). The data collected from an objective record of performance, rather than relying on self-reported behaviour, which may not reflect actual practice (Flach et al. 1998). However, observation alone cannot elicit information about the cognitive processes involved in task performance. It provides data about *what* behaviours occur but not *why* they occur. Therefore, observational methods should ideally be used alongside other methods.

2.2.2 VEHICLE MEASURES

Basic observational methods (e.g. roadside observation) enable researchers to collect crude measurements of driving behaviour, such as whether they stopped, how long they stopped and approximate travel speeds. To gain more precise measures of driving behaviour, researchers must use methods that permit them to directly record vehicle parameters such as speed, acceleration/deceleration, braking, lateral position, heading and headway. This can be done using a real vehicle (e.g. on a public road, closed course or test track) as in on-road instrumented vehicle studies, or using a driving simulator. The advantage of on-road studies is that they measure real

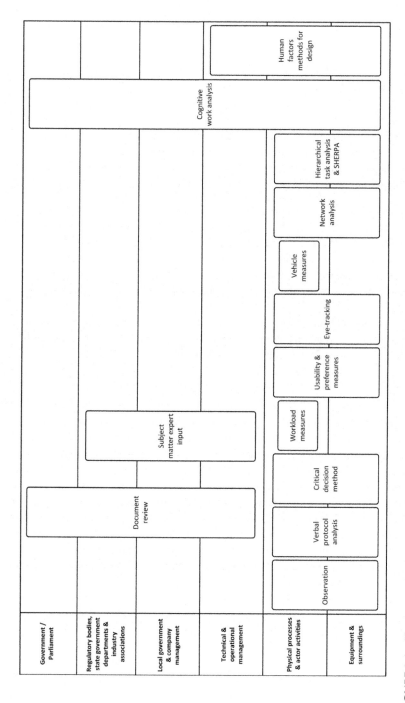

FIGURE 2.1 Key human factors methods overlaid on Rasmussen's (1997) risk management framework. SHERPA, Systematic Human Error Reduction and Prediction Approach.

BOX 2.1   OBSERVATIONS OF PEDESTRIAN BEHAVIOUR

As a part of this research programme, we used covert observational techniques to study pedestrian behaviour at rail level crossings. Considerations for selection of sites to observe included the following:

- Previous incident and near miss rates
- Sites of interest to industry subject-matter experts (e.g. due to high incident rates)
- Covering a range of sites with different features such as the type of warnings present, number of train tracks, train traffic mix (e.g. passenger vs. freight trains, express vs. stopping trains) and whether crossings were adjacent to a train station

We analysed incident data provided by the industry to understand the times of day during which most incidents or near misses occur, and then planned our observations to coincide with these times (7–9 a.m. and 2–4 p.m.). On-site observations at seven rail level crossings were undertaken in a covert manner so as not to influence user behaviour. The observer was located in a vehicle, in a signal box overlooking the crossing or on a station platform overlooking the crossing.

We also used observation to understand drivers' visual scanning behaviour at rural level crossings through video analysis and to gain familiarisation with the train driver task during cab rides (see Chapter 4).

driving, including genuine risks (e.g. the possibility of being involved in a collision or incurring a speeding ticket). It is also possible to design quasi-experimental studies, in which participants experience distinct driving conditions at different points of the drive. The corollary to this is that the road environment and traffic conditions cannot be controlled, so different participants may experience markedly different driving conditions, and confounding variables may influence behaviour in some circumstances. Certain situations may also be unethical (e.g. driving while intoxicated) or unfeasible (e.g. novel infrastructure designs) to have participants encounter in a real-world drive. For these reasons, driving simulation is commonly used to complement on-road studies, as it affords researchers greater flexibility and control in the driving scenarios that they can design.

In a typical on-road study of driver behaviour, participants drive an instrumented vehicle that is fitted with a data logging device to record vehicle speed and global positioning system (GPS) location. The data logger should be connected to the vehicle's Controller Area Network to record parameters such as the driver's interactions with the brake, accelerator and steering wheel. This enables analysis of the important aspects of driving performance, such as speed and braking. The data logger should also be connected to at least one video camera that records the road environment, to

FIGURE 2.2   Example still frame from multi-camera footage in an instrumented vehicle.

permit researchers to examine speed and braking profiles with reference to the video record (e.g. to identify whether speed changes were due to changes in traffic, road environment or infrastructure). If the setup includes only one video camera, it should film the forward road ahead of the participant. Where multiple cameras are included, these may also film the rear view, left and right exterior of the vehicle, and vehicle cockpit, including the driver (see Figure 2.2).

In a laboratory-based simulator study, vehicle measures are collected by the software system as the participant drives in the virtual environment. Measures, including speed, braking and lateral and longitudinal positions, can be extracted. These measures can be interpreted in different ways, for example, to assess the extent of driver impairment resulting from alcohol intoxication or drowsiness.

Appropriate measures should be determined during the design phase of the study, as some simulator software platforms only record parameters that are pre-specified. Other platforms record all possible parameters, but pre-specifying the variables for extraction allows for easier data processing.

A key benefit of collecting vehicle measures is the objective nature of the data. In this way, on-road and simulator studies are like observation, but offer more precise measurement (e.g. exact travel speeds). However, without using additional complementary methods such as verbal protocols, eye tracking and cognitive task analysis, the cognitive processes associated with the behaviours captured will remain unknown.

BOX 2.2   USING VEHICLE MEASURES TO UNDERSTAND DRIVER BEHAVIOUR AT RAIL LEVEL CROSSINGS

As part of this research programme, we collected vehicle measures from both an instrumented vehicle, to record drivers' behaviour when encountering existing rail level crossing designs (see Chapter 4), and a driving simulator, to test drivers' responses to novel rail level crossing designs that were developed based on our research findings (see Chapter 9).

2.2.3   EYE TRACKING

Eye tracking studies use a specialised piece of equipment called an eye tracker to record where individuals are looking while they perform a given task or view a scene. By examining which objects the participant fixated on, researchers can identify which objects in the environment captured their attention. For example, in an on-road study, are drivers focusing their visual attention on signs, road markings, pedestrians or flashing lights? Analysis of scanning patterns can also reveal which aspects of the environment the driver expects to provide the most useful information.

There are many different types of eye trackers, which vary in their functionality (Holmqvist et al. 2011). Eye trackers for studying eye movements in the real world use a combination of cameras: one to record the scene that the participant is viewing, and one or two to record the eye(s). The two camera sources are synced to produce a video of the scene that the participant saw, with an overlaid marker that represents the participant's gaze location at each point in time. Some systems are 'monocular', recording only eye, whereas other systems are 'binocular', recording both eyes. In driving studies, there is usually no meaningful advantage to using binocular recording systems, as monocular recordings provide a sufficiently accurate representation of gaze position.

Broadly categorised, there are two types of eye tracking systems that can be used for measuring eye movements in on-road studies:

1. *Head-mounted eye trackers*, which are worn by the participant (most look like a pair of spectacles with small cameras attached)
2. *Dashboard mounted eye tracking systems*, which use remote cameras to record eye movements

Dashboard systems are less intrusive than wearable systems (which can become uncomfortable after an extended period), but commercially available research-grade eye tracking systems are also less accurate at localising eye movements, which has implications for data accuracy. For instance, it may be possible to localise the general region where the driver is looking, but dashboard systems are unlikely to be accurate enough to determine whether the driver is fixating on specific signs.

Several other factors can affect the accuracy of eye tracking data, including excessive sunlight, variations in ambient lighting, and drivers wearing corrective lenses and/or eye makeup (Holmqvist et al. 2011). However, as technology continues to improve, in future these issues are likely to become less problematic.

In general, wearable eye tracking systems offer the greatest flexibility as they often have the capability to record data without being connected to a computer. Dashboard mounted systems usually require a computer connection, which limits their portability if the study involves moving the equipment between different vehicles (e.g. because participants are driving their own vehicles, or the study is using multiple vehicles in different locations).

2.2.4 VERBAL PROTOCOL ANALYSIS

The methods described thus far collect in-depth objective data, but have limited utility for uncovering the cognitive processes underlying the behaviours analysed. Verbal protocol analysis is a data collection and analysis method that can be used to elicit the cognitive and physical processes that an individual uses to perform a task. This is achieved by asking individuals to 'think aloud' while concurrently performing the task of interest, such as driving, and then analysing a transcript of these verbalisations to make 'valid inferences' from the content of discourse (Weber 1990). Example instructions that can be used to train participants in providing concurrent verbal protocols are provided in the appendix.

Verbal protocols can provide data on the user's thinking processes and content of their situation awareness while driving (Walker 2005). Accordingly, the approach is becoming popular as a way of assessing driver cognition, decision-making and situation awareness during on-road studies (e.g. Salmon et al. 2014a, Walker et al. 2011, Young et al. 2013b) and in driving simulators (e.g. Banks et al. 2014). A recent study showed that providing concurrent verbal protocols had no negative effects on driving performance, with the only effects found being positive (e.g. drivers become more aware of changes in the road environment; Salmon et al. 2016a). Although this finding is encouraging in that using such techniques in on-road studies does not place participants at increased risk for crashes, it is important to note that it may make participants more aware than they would be in everyday driving.

2.2.5 COGNITIVE TASK ANALYSIS INTERVIEWS

Cognitive task analysis techniques are used to describe the unobservable cognitive aspects of task performance. These approaches were developed to identify and describe the cognitive processes and strategies underlying decision-making, judgements and goal generation (Militello and Hutton 1998). Data collection typically relies upon retrospective interviews with experts, with structured questions that probe the key aspects of the decision-making process. A key aim of cognitive task analysis approaches is to uncover the tacit knowledge of experts, which is difficult to verbalise or to capture through traditional interview techniques (Crandall et al. 2006).

There are several methods for cognitive task analysis, with a leading method emerging being the Critical Decision Method (CDM; Klein and Armstrong 2005,

Klein et al. 1989). CDM is a semi-structured interview technique that uses structured prompts to aid recall of past events and explore factors that shape decision-making. It can be used to identify training requirements, generate training materials, provide input for the development of decision support systems and evaluate how new decision support systems impact task performance (Klein et al. 1989). It evolved from the Critical Incident Technique (Flanagan 1954) and was developed to study naturalistic decision-making strategies of experienced personnel. It has been applied within complex and dynamic domains such as firefighting, military and emergency medicine (Klein et al. 1989). It has also been used in previous transport research in aviation (Plant and Stanton 2013), maritime operations (Øvergård et al. 2015) and rail (Tichon 2007).

Probes are used within the semi-structured interview format to elicit the following types of information (see O'Hare et al. 2000 for full question set):

- *Goal specification*: What were your specific goals at the various decision points?
- *Cue identification*: What features were you looking at when you formulated your decision?
- *Expectancy*: Were you expecting to make this type of decision?
- *Conceptual model*: Are there any situations in which your decision would have turned out differently?
- *Situation assessment*: Did you use all the information available to you when formulating the decision?
- *Decision blocking – stress*: Was there any stage during the decision-making process in which you found it difficult to process and integrate the information available?
- *Analogy/generalisation*: Were you at any time reminded of previous experiences in which a similar decision was made?

Interviews conducted with CDM should be transcribed verbatim before analysis. There are several options for analysing the interview transcripts. For instance, responses can be coded using automated thematic analysis software programmes such as Leximancer™ or can be coded manually using a structured *a priori* framework, or using an exploratory emergent theme approach (Wong 2004).

BOX 2.3   USING THE CRITICAL DECISION METHOD WITH DRIVERS

As part of this research programme, we used CDM to better understand driver behaviour at rail level crossings. The outcomes of this are described in Chapter 4, along with information about how we used the method with novice as well as experienced drivers, to understand the differences in the decision-making processes of these two groups.

2.2.6 Workload

Subjective workload refers to 'the cost incurred by a human operator to achieve a particular level of performance' (Hart and Staveland 1988). It is a multidimensional construct, which emerges from the interaction between a given task and situation, as well as the skills, behaviour and perception of the operator. Measuring workload is of increasing importance within human factors as systems become more complex, imposing ever-increasing demands on workers and users. One of the most widely used measures of subjective workload is the National Aeronautics and Space Administration (NASA) Task Load Index (NASA-TLX; Hart and Staveland 1988), which was originally developed for use with pilots but has been subsequently successfully adapted for other contexts.

The NASA-TLX measures subjective workload during a specific task on multiple dimensions. Different dimensions of workload are measured through six subscales:

1. *Mental demands*: How mentally demanding was the task?
2. *Physical demands*: How physically demanding was the task?
3. *Temporal demands*: How hurried or rushed was the pace of the task?
4. *Performance*: How successful were you in accomplishing what you were asked to do?
5. *Effort*: How hard did you have to work to accomplish your level of performance?
6. *Frustration*: How insecure, discouraged, irritated, stressed and annoyed were you?

Participants rate each item on a scale from 'very low' to 'very high' except for 'Performance', which is rated from 'failure' to 'perfect' or 'poor' to 'good'. Weighted subscale scores can be combined to derive an overall measure of workload.

There are several different response formats for administering measures such as the NASA-TLX, including paper and pencil, electronically (e.g. using an iPad) or verbally (e.g. a researcher asks each question and the participant provides a verbal response, which the researcher records). The response format required will depend on the resources available, as well as the timing of administration. For example, if the NASA-TLX is administered during an on-road study, while the participant is still driving, this can only be achieved by administering it verbally, whereas in a simulator study, it would be feasible to have participants complete the NASA-TLX in between driving tasks. The NASA-TLX scale and guidance for scoring are provided in the appendix.

**BOX 2.4   MEASURING SUBJECTIVE WORKLOAD
AT RAIL LEVEL CROSSINGS**

As part of this research programme, we used the NASA-TLX within driving simulator studies to understand the differences in subjective workload between existing 'baseline' rail level crossing environments and new rail level crossing environments. See Chapter 9 for further information.

2.2.7   Usability and Subjective Preference Measures

Usability refers to how easy it is for potential users to successfully learn to use a new system with minimal effort (Davis 1989). There are various ways in which usability of a system can be evaluated.

Standard ISO/TR 16982:2002 identifies a range of usability methods and provides guidance to assist the choice of usability methods based on variables such as the constraints of the project (e.g. time available, budget available), the characteristics of users, their availability to participate in usability testing and the extent and importance of the design change (e.g. whether it is a safety-critical change). However, usability evaluation methods have tended to focus on the assessment of computer-based displays, rather than wider systems with which end users may interact in transportation domains.

The System Usability Scale (SUS; Brooke 1996) provides an efficient means to gain participants' subjective ratings of usability. It consists of a 10-item questionnaire with responses provided on a 5-point scale from 'strongly disagree' to 'strongly agree'. Scoring (described in the appendix) provides a usability score out of 100, with higher scores indicating greater usability. Like the NASA-TLX, different response formats can be used with the SUS, including paper, electronic or verbal responses.

In addition, usability assessment can involve asking open-ended questions to better ascertain users' understanding of the system and their preferences. This qualitative information can provide important depth to the quantitative data gained from scales such as the SUS and can greatly assist in directing additional design refinement activities.

**BOX 2.5   ASSESSING THE USABILITY OF RAIL
LEVEL CROSSING ENVIRONMENTS**

As part of this research programme, we adapted the SUS for use in a driving simulator study to gain subjective usability ratings of novel rail level crossing designs compared to baseline simulated environments (see Chapter 9). We also assessed usability and subjective preference of new designs using a survey methodology (see Chapter 10).

2.3   DATA COLLECTION METHODS FOR UNDERSTANDING SYSTEM PERFORMANCE

Although the methods discussed already offer an in-depth understanding of human behaviour and human interaction with technology and the immediate environment, they do not necessarily uncover more abstract aspects of the system. This leaves an explanatory gap, whereby higher levels within the system hierarchy are not addressed. To rectify this gap and ensure coverage of all stakeholders within the system, it is necessary to use other methods, including review of relevant documentation and interviews with subject-matter experts.

> ## BOX 2.6 DOCUMENT REVIEW TO UNDERSTAND
> ## THE RAIL LEVEL CROSSING SYSTEM
>
> The types of documents reviewed to inform our HTA and CWA analyses included engineering standards, road rules and legislation, policy documents and government reports.

2.3.1 DOCUMENT REVIEW AND ANALYSIS

Document review is an important familiarisation activity, especially when beginning research in an unfamiliar domain. Useful documents to source could include legislation governing system operation, technical/engineering standards, policy documents, procedures, government reports and accident investigation reports. Subject matter experts may be able to direct the research team to the most useful and credible documents to analyse.

It is recommended that the review be undertaken purposefully, with the intention of inputting into another method such as HTA or CWA. This could involve using content analysis and classifying aspects of the documents into categories relevant to the method. For example, for HTA, procedural documents could be reviewed and content classified as relating to goals, sub-goals, operations and plans (see Section 2.4.2). For CWA, technical documents and standards could be reviewed to identify the physical objects present in the system and their affordances; and policy documents and legislation could be reviewed and classified to identify the functional purposes of the system (see Section 2.4.4). This type of content analysis could be conducted using software such as NVivo (QSR International).

2.3.2 INPUT FROM SUBJECT-MATTER EXPERTS

Subject-matter or domain experts are an important source of information and input. They may participate in data collection activities such as interviews or walkthroughs, where they act as the primary data source. Experts can also provide input into the validation of analyses conducted by researchers, for example, checking and refining the HTA models derived from other data collection methods. It is important to identify relevant subject-matter experts for the domain of interest. To align with systems thinking principles, these experts should also be drawn from different levels of the system. They may include expert users or workers, managers, policy makers, system developers and technicians (Naikar 2013). Gathering subject-matter experts together to enable debate and discussion can be a valuable activity to understand where potential conflicts or areas of complexity might exist (Flach et al. 1998). This could be achieved via a group interview or focus group session.

However, it is acknowledged that domain experts can be difficult to access (Potter et al. 1998), so approaching them for input may require some planning as to how much input is required and when it may be most valuable. Potentially, efficiencies can be gained by applying techniques such as documentation review and observation,

> **BOX 2.7   SUBJECT-MATTER EXPERT INPUT FOR**
> **UNDERSTANDING THE RAIL LEVEL CROSSING SYSTEM**
>
> In developing our CWA outputs, we gained input from a small amount of subject-matter experts initially, using semi-structured interviews, and later used a focus group with 11 domain experts to validate the Work Domain Analysis model (see Chapter 5).

and generating initial analyses, followed by engagement with subject-matter experts for validation of the analysis.

2.4   SYSTEMS-FOCUSSED ANALYSIS METHODS

Data collected using the methods described in Sections 2.2 and 2.3 can be analysed using a variety of techniques, including simple descriptive analyses and inferential statistics to compare behaviour and performance between conditions. Although these methods are useful to provide an initial understanding and overview of the raw data, a key advantage of many of these data collection methods is the ability to analyse the data in greater depth using additional human factors analysis techniques that offer specific insights about system functioning.

2.4.1   NETWORK ANALYSIS

Network analysis can be used to understand the relationships between tasks, social actors (humans and technology) and knowledge concepts. In this research programme, we have generally used network analysis in relation to the latter, to focus on the situation awareness of users at rail level crossings. Here, the data input is the verbal protocol transcripts.

Situation awareness networks illustrate the concepts verbalised by participants and the relationships between them, to provide a detailed picture of what participants' situation awareness comprised at key points. Within situation awareness networks, the nodes represent pieces of information or concepts relevant to situation awareness (e.g. speed, lights).

Situation awareness networks can be constructed manually or via software tools such as Leximancer, which uses text representations of natural language to interrogate verbal transcripts and identify themes, concepts and the relationships between them. The software has previously been used for situation awareness network construction and analysis in various on-road studies (Salmon et al. 2013a, 2013d, 2014a, 2014b, Walker et al. 2011). An important strength of this approach is that it provides a reliable, repeatable process for constructing situation awareness networks.

A further strength of using networks to describe participant situation awareness is the ability to analyse them in various ways using network analysis metrics. Various metrics are used to analyse situation awareness networks. Metrics that may have relevance include the following:

- *Network density*: This metric represents the level of interconnectivity of the network in terms of relationships between nodes. Density is expressed as a value between 0 and 1, with 0 representing a network with no connections between nodes, and 1 representing a network in which every node is connected to every other concept (Kakimoto et al. 2006).
- *Sociometric status*: This metric provides a measure of how 'busy' a node is relative to the total number of nodes within the network under analysis (Houghton et al. 2006). Nodes with sociometric status values greater than the mean sociometric status value plus one standard deviation may be designated as 'key' (i.e. most connected) nodes in the social and situation awareness networks.
- *Centrality*: This metric measures the standing of a node within a network in terms of its distance from other nodes in the network (Houghton et al. 2006). A 'central' node is relatively close to all other nodes in the network in terms of connections. That is, an interaction with other nodes in the network is achieved through the lowest number of connections.

The use of these metrics enables conclusions to be made regarding the structure of situation awareness. In an on-road study context, for example, this allows conclusions to be made regarding:

- The most important pieces of information being used when negotiating road environments (sociometric status, centrality).
- Differences in the connectedness of information when drivers negotiate different types of road environments, such as intersections versus rail level crossings (e.g. network density: does situation awareness become harder to attain and is the driver required to use more information?).
- Instances where important pieces of information (e.g. speed reductions) are not well integrated in drivers' understanding of the situation (sociometric status, centrality).

Analysis of data via network analysis metrics is normally supported through a network analysis software tool such as Agna. This involves importing the network data into Agna in the form of a matrix of the concepts (e.g. car, traffic lights, pedestrian, speed) and the relationships between them (e.g. 'car' was mentioned with 'speed' seven times; 'car' was mentioned with 'pedestrian' once). The network metrics are then calculated automatically by the software tool by selecting the appropriate metrics.

BOX 2.8   USING NETWORK ANALYSIS TO UNDERSTAND SITUATION AWARENESS AT RAIL LEVEL CROSSINGS

In this research programme, we applied network analysis to the transcripts of verbal protocols obtained during on-road studies to understand driver situation awareness at rail level crossings (see Chapter 4).

2.4.2 HIERARCHICAL TASK ANALYSIS

Hierarchical Task Analysis (HTA) is a method for understanding the hierarchical decomposition of system goals, based on a desired objective or end state (Annett and Stanton 1998). Although HTA originated from scientific management methods and is often employed to decompose the tasks of individual operators (Stanton 2006), it is ultimately a systems method, enabling analysts to decompose the whole system in terms of its goals, sub-goals and requisite activities. It provides a normative model of functioning, that is, it describes what should be done to achieve the overall goals of the system.

The HTA method was developed in response to a need to better understand cognitive tasks brought about by the changing nature of industrial work processes during the 1950s and 1960s (Annett 2004). At the time, the focus on cognition as well as physical work made it unique, and it subsequently became arguably the most popular of all human factors methods (Stanton 2006).

HTA works by decomposing systems into a hierarchy of goals, sub-ordinate goals, operations and plans; it focuses on 'what an operator … is required to do, in terms of actions and/or cognitive processes to achieve a system goal' (Kirwan and Ainsworth 1992). It is important to note here that an 'operator' may be a human or a technological operator (e.g. system artefacts such as equipment, devices and interfaces). HTA outputs therefore specify the overall goal of a system, the sub-goals to be undertaken to achieve this goal, the operations required to achieve each of the sub-goals specified and the plans that are used to ensure that the goals are achieved. The plans component of HTA is especially important as they specify the sequence, and under what conditions, different sub-goals must be achieved to satisfy the requirements of a super-ordinate goal.

HTA has been widely used in several domains, including the process control and power generation industries, the military (Ainsworth and Marshall 1998, Kirwan and Ainsworth 1992) and transport, including aviation (Stanton et al. 2009) and bus driving (Salmon et al. 2011). It has also been adapted for use in a range of human factors applications, including training (Shepherd 2002), design (Lim and Long 1994), error and risk analysis (Baber and Stanton 1994) and the identification and assessment of team skills (Annett et al. 2000).

**BOX 2.9   HIERARCHICAL TASK ANALYSIS
FOR RAIL LEVEL CROSSING SYSTEMS**

We used HTA to understand the rail level crossing system, in terms of the goals, sub-goals and operations of the key actors, including road users, train drivers, the train and the rail level crossing infrastructure. See Chapter 5 for further information.

2.4.3 Systematic Human Error Reduction and Prediction Approach

Systematic Human Error Reduction and Prediction Approach (SHERPA) is a human error identification approach, originally designed to assist those working in the process industries (e.g. conventional and nuclear power generation, petrochemical processing, oil and gas extraction and power distribution; Embrey 1986) to better understand error potential. The domains of application have broadened in recent years, to include areas such as aviation (Harris et al. 2005) and health care (Lane et al. 2006, Phipps et al. 2008).

SHERPA is based on normative models of task performance, such as HTA. Where HTA has been completed at the systems level as discussed previously (i.e. where 'operators' have been defined as including both human and technological actors), SHERPA analysis enables an understanding of potential failure across all aspects of the system. SHERPA uses a taxonomy to classify different types of potential errors. The taxonomy is based on the following five task types:

1. Action (e.g. pressing a button, engaging a piece of equipment, opening a door)
2. Retrieval (e.g. retrieving information from a display or manual)
3. Checking (e.g. conducting a check for signage)
4. Selection (e.g. choosing one alternative over another)
5. Information communication (e.g. exchanging information through verbal or non-verbal means)

The outcome of a SHERPA analysis is a set of credible errors that can be prioritised based on probability and criticality, with concomitant error reduction strategies or interventions.

**BOX 2.10   HUMAN ERROR IDENTIFICATION
FOR RAIL LEVEL CROSSINGS**

In this research programme, we used SHERPA initially to understand the potential errors associated with the current system (see Chapter 5) and then subsequently within a design evaluation and refinement process to assess how new designs would address the existing errors or introduce new errors (see Chapter 8).

2.4.4 Cognitive Work Analysis

Cognitive Work Analysis (CWA) is a framework developed to model complex socio-technical work systems (Jenkins et al. 2009, Rasmussen et al. 1990, Vicente 1999). However, it can also be used beyond traditional work systems, as recent applications in road transport and other areas have demonstrated (e.g. Cornelissen et al. 2012).

BOX 2.11   COGNITIVE WORK ANALYSIS TO UNDERSTAND THE RAIL LEVEL CROSSING SYSTEM

CWA was an integral method used in this research project. See Chapter 5 for information about the analysis and Chapter 6 for details on how we used insights from the CWA to inform the design of novel crossing designs.

The CWA framework is focussed on identifying the constraints on system functioning; it generates representations of the system that enable the analyst to understand how activity could proceed within a given system. The focus on constraints separates the technique from other analysis approaches that are descriptive (i.e. describing how activities *are* conducted), or normative such as HTA (i.e. prescribing how it *should* be conducted).

The framework comprises five phases of analysis that are selected based upon the aims of the analysis and/or design process. The phases commence by modelling the environmental constraints on the domain, with subsequent phases progressively narrowing the focus to consider constraints associated with tasks, strategies, allocation of functions and cognitive skills required by workers or users interacting within the domain (Vicente 1999). The five phases are as follows:

1. Work Domain Analysis
2. Control Task Analysis
3. Strategies Analysis
4. Social Organisation and Cooperation Analysis
5. Worker Competencies Analysis

The system descriptions provided by each phase can then be used to address specific research and design aims. For example, Work Domain Analysis is commonly used to support interface design and evaluation purposes, but it can also be used to inform training design and evaluation.

2.5   HUMAN FACTORS DESIGN METHODS

Several of the methods already discussed are used by human factors practitioners and researchers to inform design by providing analytical findings or insights, which can be incorporated into the system redesign process. However, there are a set of methods that are specifically intended to assist the creative process of design and that align with the systems thinking and sociotechnical systems theory approaches described in Chapter 1. A selection of these methods is introduced here.

2.5.1   SCENARIOS AND STORIES

Scenarios are narratives describing 'use situations' relevant to the design process. They can be used to communicate important information about user characteristics and task context efficiently and effectively to those involved in design (Carroll 2002). Narratives help to promote an understanding of user goals, experiences and challenges to prompt design solution ideas. They can also be used to evaluate proposed designs, by considering how the scenario might be different if a new design was implemented. Systems-focussed analysis methods, particularly CWA, can provide many inputs to scenarios, including user goals, the presence of conflicting goals, different types of situations and circumstances, different strategies that can be employed, and different competencies of users.

Stories, while similar to scenarios, are different in that they represent real-life cases describing people's interactions within the system and can be used to demonstrate a particular finding or perspective that is illustrative to discuss. They can be used for communicating information uncovered during the analysis in a concrete, specific way (Erickson 1995). They can also assist to promote empathy with users or other stakeholders within the system.

2.5.2 PERSONAS

Personas are descriptions of people (real or fictional) within the system. They provide information about the person's circumstances, capabilities and limitations, motivations, values and so on to assist designers to develop empathy with various users and stakeholders of the system. Empathy is important for achieving user-centred designs that align with sociotechnical systems theory values.

2.5.3 INSPIRATION CARDS

Inspiration cards and card sorting techniques are commonly used within participatory design activities. Tangible materials, such as cards, can provide a means of engaging participants and encouraging their physical, rather than only cognitive, involvement in the design process (Halskov and Dalsgård 2006).

Some sets of cards have been developed to inspire design ideas. For example, cards can be used in design games such as those described by Brandt and Messerter (2004). In the User Game, design participants use Moment cards (reflecting short videos of naturalistic activities) and Sign cards (with printed concepts such as 'despair', 'pace', 'vibrant', 'closeness' and 'zones') to build and annotate stories relating to the design question. The game follows an approach similar to a crossword, where one story will be outlined with cards, which then is added to by the next player – intersecting at one point in the story. This builds a web of interconnected stories that envisage potential user experience within a system.

Another card-based design method is the Design with Intent Toolkit (Lockton et al. 2010), created to provide design patterns that can be used in design for behaviour change. The toolkit provides cards based on eight different lenses or perspectives on design, both environmental and cognitive. For example, in the Architectural Lens, a card titled 'Positioning' asks the designer to consider 'Can you rearrange things so people interact with them in the locations you want them to?'

The toolkit provides information about each lens and guidance on how the cards can be used to generate design ideas. Novel uses of the cards are also encouraged.

Furthermore, designers may create their own card sets relevant to the design questions being addressed.

2.5.4 ASSUMPTION CRUSHING

This tool outlines a process for considering assumptions associated with the existing system, then challenging these assumptions and creating alternative statements that change the boundaries of the design space (Imber 2012). This enables the design team to transcend traditional design solutions and promotes innovation. The alternative statements developed can be used as a basis to brainstorm design ideas.

2.5.5 METAPHORS AND ANALOGIES

Metaphors and analogies can assist designers to take inspiration from an area or domain that is similar but has some difference, and apply this in design. Metaphors provide a means of seeing familiar things in a new way. For example, Norman (2007) describes how the metaphor of the horse and rider can be used to explore approaches to designing vehicle automation.

**BOX 2.12   HUMAN FACTORS DESIGN METHODS TO
GENERATE NOVEL RAIL LEVEL CROSSING DESIGNS**

In our research programme, we used these design methods and tools in a participatory design process to generate new designs for rail level crossings. See Chapter 6 for more information about this process.

2.6 SUMMARY

This chapter has outlined the key methods and approaches that were utilised in this research programme. In the remainder of this book, we will explain how these various perspectives on understanding human and system functioning were integrated and used to inform the analysis, design and evaluation of rail level crossings.

3 An Integrated Framework for Transport Analysis and Design

3.1 INTRODUCTION

In this chapter, we will describe how we integrated the methods described in Chapter 2 to draw upon the strengths of each within an overall systems thinking approach. Although the discipline of human factors is focussed on applied outcomes, relatively few research programmes have involved comprehensive data collection, analysis, design and formal evaluation under the umbrella of a single research project. Instead, human factors research by necessity is often limited to a piecemeal, reductionist approach involving exploration of a single issue, highlighting primary findings and providing specific recommendations for change. This may be because the findings are needed quickly (e.g. to input within a design process) or because of the limitations of funding available for human factors research. We had the privileged opportunity with a multi-year programme of research to go beyond exploring the problem, to a process of redesign and evaluation. The aim of the research was to recommend practical outcomes that could be implemented to improve safety, underpinned by a systems thinking framework. Although the framework was developed to improve rail level crossing safety, it was designed to be generic to enable modification for research on other transport safety issues.

3.2 A RESEARCH PROGRAMME UNDERPINNED BY SOCIOTECHNICAL SYSTEMS THEORY

As noted in Chapter 1, sociotechnical systems theory underpins the approach taken in the research described in this book. Emerging in the 1950s and since applied to organisational design for many decades (Mumford 2006), sociotechnical systems theory is strongly aligned with systems theory and underpinned by notions of industrial democracy, participatory design and humanistic values. In short, the sociotechnical systems theory approach aims to design organisations and systems that have the capacity to adapt and respond to changes and disturbances in the environment.

Sociotechnical systems theory approaches are traditionally adopted within an action research paradigm, responding to a known problem that has been identified by stakeholders, in some cases through experiential and informal knowledge. Consequently, they often integrate problem exploration within the design generation stage, without a dedicated background of data collection and analysis. For this research programme, it was clear from reviewing the literature (e.g. Edquist et al. 2009) and discussions with stakeholders that there was not an adequate understanding of the problem from the outset. This necessitated a comprehensive data collection programme to help understand the aspects of the problem and to provide a robust basis for Cognitive Work Analysis (CWA) and the other systems-based analysis methods applied.

As described in Chapter 2, CWA is a sociotechnical systems framework: it is concerned with how individuals and groups (*the socio*) and technology or artefacts (*the technical*) interact to achieve the goals of the system. However, CWA has previously been criticised for failing to contribute directly to design processes (Jenkins et al. 2010, Lintern 2005, Mendoza et al. 2011). Thus, one of the innovations of this research programme was to develop a design process for use with CWA, to demonstrate how the phases of CWA can be used within a larger sociotechnical systems theory-based framework for analysis, design and evaluation. The outcome of this, the CWA Design Toolkit (CWA-DT), is described in Chapter 6.

To inform the design of our research programme, we looked to the principles and values of sociotechnical design that have evolved over many years of action research implementing innovations in organisations (e.g. Cherns 1976, Clegg 2000, Davis 1982, Walker et al. 2009b). These principles intend to support the design of sociotechnical systems that exhibit adaptive capacity. Two sets of principles have been defined (Read et al. 2015b):

1. Process principles: What an sociotechnical systems theory-based design process should encompass
2. Content principles: What the outcome of the process (the design itself) should encompass (see Chapter 6)

In Table 3.1, we have defined the process principles and described how our research programme was designed to address each principle.

In addition, the values of sociotechnical systems theory underpin the design process and should also be represented in the outcomes of the design process (Cherns 1987). Sociotechnical systems theory values will be familiar to human factors professionals as these have permeated the discipline generally over the years. The values include the following:

- *Humans as assets*: Rather than characterising humans as unpredictable, error-prone and the cause of problems in otherwise well-designed technological systems, sociotechnical systems theory acknowledges that no technical system is perfect and that humans are assets capable of identifying the need for change, of learning and adapting, and of effective problem solving (Clegg 2000, Norros 2014).
- *Technology as a tool to assist humans*: Technology should be viewed as a tool to assist people to meet their goals, rather than an end in its own right (Clegg 2000, Norros 2014). Frequently, technical solutions are proposed as a panacea to a problem, often as a replacement for humans, with little or no

TABLE 3.1

How the Sociotechnical Systems Theory Process Principles Were Considered in the Design of the Research Programme

STS Process Principle	Description	Application to the Research Programme
Adoption of appropriate design process	Design approaches, methods and techniques must be matched to the fundamental nature of the problem and its environment (Walker et al. 2009b). In addition, the systems used to undertake design also need to be designed. Sociotechnical thinking, ideas and principles are applicable to design systems (Clegg 2000). Importantly, the process of design or re-design needs to be compatible with its objectives (Davis 1982).	*Overall approach*: We selected the STS philosophy to underpin the entire research programme, given the nature of rail level crossings as complex, open systems (Read et al. 2013). We applied this philosophy and its core values to design our research approach as well as to drive the design aspects of the research.
Provision of resources and support	Resources and support are required for design. This includes expertise in how to adopt a more holistic and systemic view. The process needs to be supported by appropriate methods and tools, working principles, theoretical understanding and frameworks (Clegg 2000).	*Overall approach*: We conducted the project with 4 years of funding and support from industry partners, including access to data, information and expertise. The research was led by experts in human factors and systems thinking and supported by appropriate methods and tools, with a focus on CWA.
Context/problem analysis	Design should be appropriate to the specific context, rather than simply involving the uptake of imported or copied solutions (Davis 1982). Design decisions must also consider competing demands and opportunity costs (Clegg 2000).	*Data collection*: We were careful in selecting data collection methods to ensure that different rail level crossing contexts were considered (e.g. urban vs. rural environments; active vs. passive crossings) and that the perspectives of different types of end users were gained. This provided an in-depth view of the problem. *Design*: We chose to use a participatory design process focussed on developing novel ideas for improving safety and system performance, involving diverse stakeholders and subject-matter experts, to optimise the balancing of competing demands and values in making design decisions.

(Continued)

TABLE 3.1 (*Continued*)

How the Sociotechnical Systems Theory Process Principles Were Considered in the Design of the Research Programme

STS Process Principle	Description	Application to the Research Programme
Constraints are questioned	Constraints used to criticise or rebuff novel design ideas (e.g. cost, time) should be questioned, to avoid prematurely closing off options (Cherns 1976).	*Systems analysis*: We selected CWA as an analysis framework to enable us to identify the key constraints on the functioning of level crossing systems and their re-design. This enabled the questioning and challenging of existing constraints within the design process, using design tools and techniques such as 'assumption crushing'.
Representation of interconnectedness of system elements	All aspects of a system are interconnected and none should take logical precedence over the other. System elements should be designed jointly, and the impact of changes on aspects such as roles and structures must be considered (Davis 1982). Efforts should be made to trace through unintended effects of a design and to facilitate evaluation after implementation (Clegg 2000).	*Systems analysis*: The selection of systems-based analysis frameworks and methods, such as CWA and HTA, enabled us to understand and represent the interrelationships between system components. *Evaluation*: We used the systems analysis outputs to evaluate the proposed design concepts to understand the effects on the wider system.
User participation	The design process must be compatible with its objectives. If the objective is to create a system capable of adaptation and self-modification, then individuals must be provided with the opportunity to participate in the design process (Cherns 1976). Successful implementation depends upon the ownership of the design. Participation is essential for those who will be responsible for achieving successful operation (i.e. managers, supervisors and workers; Davis 1982)	*Design*: We adopted a participatory design process to promote ownership of the final outcomes by key stakeholders. We also invited user representatives to participate in the design process.
Multidisciplinary participation and learning	Design involves multidisciplinary education and should draw upon those with expertise in both the social and technical domains (Clegg 2000).	*Design*: The participatory design process adopted in the research programme provided a collaborative environment to support multidisciplinary learning among the researchers, stakeholders and other subject-matter experts.

(*Continued*)

TABLE 3.1 (*Continued*)

How the Sociotechnical Systems Theory Process Principles Were Considered in the Design of the Research Programme

STS Process Principle	Description	Application to the Research Programme
Adoption of agreed values and purposes	Those participating in the design process must reveal assumptions and reach decisions by consensus (Cherns 1987).	*Design*: We used a participatory design process that encouraged a consensus-based approach to decision-making.
Documentation of how choices constrain subsequent choices	Design involves making choices between alternatives. The choices made early in the design process are not independent and constrain later choices available to designers (Clegg 2000).	*Design*: We used a structured process to plan the design process meaning that the scope was documented from the beginning. This enabled agreement to occur about the initial constraints imposed on the process.
Design driven by good solutions – not fashion	Design occurs within a social context and is subject to social movements and trends, fads and fashions (Clegg 2000).	*Design*: Our participatory design process encouraged stakeholders to consider the best solutions rather than to follow the current trends. This was promoted by reference to STS values and design tools to focus on the needs of end users.
Joint design of social and technical elements	The design process requires a joint evaluation of the impact of the technical system design on the social system, and the impact of the social system on the operation of technical system. Optimising outcomes requires the joint design of the technical and social aspects of the system (Davis 1982). Design decisions should be reached for both technical and social reasons (Cherns 1987).	*Analysis*: Through the application of systems-based analysis methods, we gained an understanding of the interrelationships between the social and technical aspects of existing rail level crossing systems. *Design*: We used the knowledge from the analysis as well as design approaches and tools aimed to support joint design. In addition, we used the STS content principles to conduct an initial high-level evaluation of proposed designs to determine the extent to which joint design was achieved and to recommend appropriate refinements to designs.
Political debate	System design involves political processes, and as such mechanisms are needed to handle political debates and discussions (Clegg 2000).	*Design*: Our participatory design process was structured to encourage debate and discussion between stakeholders with different perspectives and responsibilities for rail level crossing safety. This was especially encouraged during the design refinement process.

(Continued)

TABLE 3.1 (*Continued*)
How the Sociotechnical Systems Theory Process Principles Were Considered in the Design of the Research Programme

STS Process Principle	Description	Application to the Research Programme
Iteration and planning for ongoing evaluation and re-design	Design is an iterative and extended social process. As soon as a design is implemented, its consequences indicate the need for redesign (Clegg 2000, Davis 1982). User requirements are not static; new needs will be discovered after implementation and new types of users may also become apparent. Systems must therefore be designed for adaptability and change, including the provision of structures and mechanisms to promote ongoing evaluation, feedback and re-design (Clegg 2000, Walker et al. 2009b).	*Design*: Our design process was intended to support the generation of designs that would be adaptable to changing circumstances and identify and support re-design opportunities.
Design and planning for the transition period	To avoid the common issues in implementing a new design or re-design, the components required for successful implementation need to be considered within the design process. A transitional phase of operation may be required to bridge the gap between the existing operations and the new design (Davis 1982).	*Evaluation*: Although our research programme did not specifically focus on the transition period, possible emergent risks were examined in our evaluation processes. In addition, the recommendations made, especially around improving data systems, could assist in managing this period and monitoring for emergent issues.

STS, sociotechnical systems theory.

consideration of the goals of people's work or the social system required to make the technology work (Clegg 2000).

- *The need to promote quality of life*: People cannot be considered as simply machines or extensions of machines (Robinson 1982). They should be provided with quality work or tasks that are challenging, have variety, include the scope for decision-making and choice, facilitate ongoing learning, incorporate social support and recognition of people's work, have social relevance and lead to a desirable future (Cherns 1976, 1987).
- *The need to respect individual differences in design*: Individuals have varied needs and preferences. Some may prefer high levels of autonomy and control; others may prefer more prescriptive guidance. The design process needs to recognise and respect differences and work towards achieving a flexible design that incorporates different preferences, acknowledging that meeting all needs may not always be possible (Cherns 1976, 1987).

- *The need to demonstrate responsibility to all stakeholders*: The effects of design choices on all stakeholders (e.g. users, manufacturers, unions, industry bodies, government bodies and the wider community) should be considered. Potential negative effects to be considered include physical damage or injury, economic loss and social and environmental harms (Cherns 1987). Impacts should be considered throughout all stages of the system life cycle from design through to de-commissioning.

We applied these values throughout the research programme. For example, we considered our research participants and project stakeholders to be assets, recognising the value of their input and expertise. Further, we used a team-based approach for the research that drew on the strengths, expertise and working preferences of team members. We worked as a distributed team across three geographical locations and used technology to collaborate. However, we found that times when team members were co-located for important tasks, such as generating initial CWA analyses, were most worthwhile due to the ease of interaction and opportunities provided for learning from one another through working collaboratively. Quality of working life was strong, with the research work being challenging, having variety and contributing to an important real world problem.

3.3 THE RESEARCH FRAMEWORK

The research process consisted of four distinct phases, as shown in Figure 3.1, which moved from data collection to understand the existing system, to whole systems analysis, to design generation and finally to evaluation and testing of the novel designs.

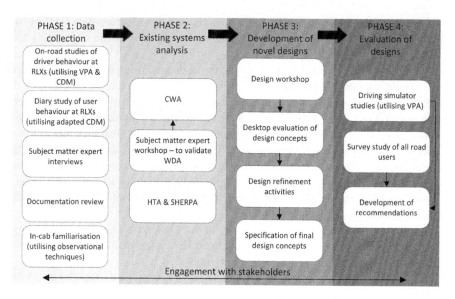

FIGURE 3.1 The phases of the research programme. RLX, rail level crossing; WDA, work domain analysis; VPA, verbal protocol analysis; CDM, critical decision method; HTA, hierarchical task analysis; SHERPA, systematic human error reduction and prediction approach.

The ultimate outcome was a set of recommendations around potential changes to rail level crossings to improve safety, but there were other innovative outcomes along the way, including the generation of the CWA-DT, which was adopted predominantly in Phase 3 to generate novel systems-focussed designs. As shown across the bottom of the diagram, we engaged with stakeholders throughout the process, both through interactions with the project steering committee and with wider stakeholders such as subject-matter experts for various distinct activities.

Each phase of the research programme is described in Sections 3.3.1 through 3.3.4.

3.3.1 PHASE 1 – DATA COLLECTION

The data collection focussed on understanding the existing rail level crossing system from multiple perspectives using multiple methods. Semi-naturalistic on-road studies of driver behaviour were conducted, combining the collection of:

- Vehicle measures and video recordings of the driving environment to understand objective driver behaviour.
- Eye tracking to understand where visual attention was directed within the road and in-vehicle environments.
- Verbal protocol analysis (VPA), analysed using network metrics, to understand drivers' situation awareness when traversing crossings.
- Cognitive task analysis interviews using the Critical Decision Method (CDM) approach to explore driver decision-making around whether to 'stop or go' at the crossing.

This mix of methods included both objective and subjective data, telling us both what drivers were *doing* and what they were *thinking* while negotiating the crossing.

Although these on-road studies provided in-depth data, being largely naturalistic it was clearly not possible to control the environment or manipulate driver exposure to specific conditions such as whether a train was present during participants' encounters with rail level crossings. This approach therefore yielded more data on situations when there was no train present, rather than situations in which the road user had to make a decision involving an approaching train. To address this, we also conducted a self-report diary study to increase the data available on decision-making at rail level crossings, using an adapted CDM approach. Importantly, this study also provided insight into the decision-making processes of cyclist, pedestrian and motorcyclist users at rail level crossings.

Finally, to gain a broader understanding of the whole system and its functioning, we interviewed subject-matter experts, analysed key system documentation and rode in train cabs during normal passenger operations.

See Chapter 2 for an overview of the data collection methods and Chapter 4 for more information about their application in this research.

3.3.2 PHASE 2 – EXISTING SYSTEMS ANALYSIS

Phase 2 of the research programme involved generating systems models of rail level crossings, using the data collected during Phase 1. Although the data collection activities alone provided a number of key findings and insights into specific aspects of user behaviour, it was not until the data were synthesised through a systems thinking lens that a deeper understanding could be gained. The formative analysis provided through CWA outputs yielded pivotal information around the system constraints that affect the functioning of rail level crossing systems and activity conducted within them.

In addition to analysing the data collected in Phase 1, we engaged with subject-matter experts from the rail and road industries to validate the Work Domain Analysis models generated for rail level crossings. This was an important step as the Work Domain Analysis underpins the latter phases of CWA. It also provided an opportunity to engage with stakeholders and familiarise them with systems thinking ideas and methods.

In addition to CWA, Hierarchical Task Analysis (HTA) and the Systematic Human Error Reduction and Prediction Approach (SHERPA) were applied using an explicitly systems-focussed perspective. Instead of a traditional HTA, which focusses on the goals of individuals (workers, operators or users), this HTA considered all system actors (both humans and technologies) and provided a decomposition of the system goals, sub-goals, operations and plans. In this way, the SHERPA taxonomy could be used to identify potential errors or failures across the entire system, rather than only those associated with human behaviour. Conducting HTA and SHERPA presented an interesting comparison to the CWA and produced a normative standard against which designs could later be evaluated.

See Chapter 2 for an overview of CWA, HTA and SHERPA methods, and Chapter 5 for more information about their application in this research.

3.3.3 PHASE 3 – DEVELOPMENT OF NOVEL DESIGNS

One of the major challenges for the research programme was bridging the gap between systems-based analyses (especially CWA) and design, to ultimately create designs that optimised system performance and safety. To assist with this, we developed the CWA-DT (Read 2015, Read et al. 2016a), which provides a process for taking the outputs of system analyses and applying them within a participatory design paradigm. A toolkit approach was chosen for the CWA-DT due to the wide range of design purposes and application domains within which CWA-based design might occur. Although there are several design toolkits available, no formal toolkits were identified as being developed for CWA, or for sociotechnical systems theory-based design.

The CWA-DT supports an iterative rather than sequential design process. Consistent with the notion of CWA as a framework rather than a prescriptive method, users of the CWA-DT have flexibility in which design tools they select, based on the purposes and scope of the design activity. The fundamental aspects of the approach include the following:

- The identification of 'insights' from systems-based analyses, including using structured prompt questions applied to the analysis outputs.
- Support for planning the design process, using templates to document the design brief and evaluation criteria that will be used to measure success, based on the analysis outputs.
- Guidance for selecting and using human factors design tools, such as scenarios, stories, personas, inspiration cards, assumption crushing and metaphors and analogies (see Chapter 2), and using them within a participatory design setting (see Box 3.1 for further information about participatory design).
- Guidance for high-level evaluation of proposed designs using the outputs of the systems-based models and sociotechnical systems theory principles, with the findings feeding into a design refinement process.

We used insights gained from Phases 1 and 2 of the research programme as the basis for a participatory design process involving road and rail industry and government stakeholders, as well as road user group representatives. The design process involved idea generation workshops, which used tools to encourage lateral and creative thinking, in which initial design concepts were generated. These concepts were then subjected to a desktop evaluation process by the researchers using CWA, HTA and SHERPA. The evaluation results were provided back to the design participants during an additional workshop. The evaluation outputs were used to make refinements and prioritise the designs to determine a shortlist for further, more detailed human factors evaluation.

See Chapters 6 through 8 for details of the design process and outcomes.

BOX 3.1   PARTICIPATORY DESIGN

The participatory approach to design is strongly values driven and views the benefit of user involvement as stemming from the following three motivations (Gregory 2003):

1. Applying the knowledge of users to create better designs
2. Taking users on the design journey, so they have a better understanding of reasons for decisions and will be less resistant to change
3. Respecting the rights of humans to have a say in decisions that will affect their interactions within the system

3.3.4 Phase 4 – Evaluation of Designs

The shortlisted design concepts generated using the CWA-DT, along with two additional concepts designed by the research team, underwent formal evaluation and testing to understand the potential benefits of their implementation.

The principal approach we used for this phase of the research programme was driving simulation. Simulator studies provide an ideal platform for initial evaluation of new design concepts, as it is possible to have users experience the new system design without having to build prototypes in the real world. It is also possible to experimentally control exposure to trains and to control the driving environment such that the only difference between rail level crossings is the type of design encountered. This minimises potential for confounding factors between different designs, which are likely to occur in field observations. Simulation also enabled us to test the designs in situations that might be unethical in on-road studies, such as deliberately distracting drivers from the task of searching for trains and exposing participants to failed warning conditions (see Chapter 9).

We used a series of driving simulator studies to evaluate how the proposed designs influenced driver behaviour. A fundamental aspect of the evaluations was that each new design was evaluated as a complete system, most of which included multiple new attributes, rather than evaluating each possible component change in isolation. This was done deliberately so that the evaluation reflected systems principles and captured the interaction of all new elements of the system, rather than how they function individually. We used a combination of objective and subjective methods to understand driving performance within the simulator studies, including collection of vehicle measures, VPA and measures of usability and workload.

Like the on-road studies conducted in Phase 1, driving simulation was limited to testing driver responses to the proposed design concepts. To gain the perspectives of users other than drivers, we conducted an online survey study where users watched video simulations of approaching and traversing each design, and then rated their perceptions of key measures of success: safety, efficiency, compliance and overall preference (see Chapter 10).

Overall, the evaluation process highlighted the aspects of the proposed designs that were likely to support desirable system functioning, as well as the aspects that did not appear to demonstrate clear benefits compared with the existing standards. In addition, issues of potential emergent risks were highlighted. The evaluation findings supported the development of recommendations provided to the industry for improving level crossing safety.

See Chapters 9 and 10 for details of the evaluation process and outcomes.

3.4 SUMMARY

In this chapter, we have outlined the approach taken to understand and model the existing level crossing system, develop novel designs and evaluate the potential effectiveness of the designs using human factors and systems thinking approaches.

The description has demonstrated how we combined a range of human factors methods for analysis, design and evaluation within a range of contexts, incorporating such broad approaches such as semi-naturalistic studies, participatory design and laboratory studies, all underpinned by a systems thinking framework. Such diversity in approaches is a key strength of the human factors discipline, where methods are used, combined and adapted as required to meet the needs of the research question at hand.

The remainder of this book is structured around the phases of the research programme. The chapters provide additional details about the methods we used, the key findings obtained and the insights we gained that may be valuable to researchers and practitioners undertaking similar endeavours.

Section II

Rail Level Crossing Data
Collection and Analysis

4 Understanding the Factors Influencing User Behaviour

With Contributions from:
Ashleigh Filtness, Kristie Young,
Christine Mulvihill, Casey Rampollard
and Nebojsa Tomasevic

4.1 INTRODUCTION

This chapter will describe Phase 1 of the research programme, which involved a series of data collection activities aimed at understanding how road users interact with existing rail level crossing systems. Road user behaviour was investigated via on-road studies, observations, interviews and surveys. Train driver experiences were explored during cab rides through urban and rural areas. Data collection methods were deliberately structured so that the data could inform Cognitive Work Analysis (Jenkins et al. 2009, Vicente 1999) and Hierarchical Task Analysis (Stanton 2006), which respectively provide formative and normative descriptions of system functioning (see Chapter 5).

The impetus for these data collection activities arose from a review of the existing literature on human factors issues in level crossing safety (Edquist et al. 2009). The most effective existing solutions to improve safety, namely, grade separation and upgrading from passive signage to boom gates, are generally considered cost prohibitive (Cairney et al. 2002, Wigglesworth and Uber 1991), particularly given there are approximately 8,838 public road level crossings in Australia and 67% of these have passive controls (i.e. signage) only (RISSB 2009). However, many lower cost interventions are not evidence based and/or have not been appropriately evaluated (Edquist et al. 2009). Although a substantive body of research exists examining road user behaviour at rail level crossings, most studies have relatively narrow scope (e.g. focussing on a single road user group) and are not been conducted in a manner consistent with systems thinking approaches (Read et al. 2013). These issues highlighted the need for a more comprehensive, in-depth application of human factors methods to understand behaviour and the factors influencing behaviour at rail level crossings.

A multifaceted data collection approach was adopted to meet the requirements for systems-based analysis. Whereas most previous research has focussed on car drivers in isolation (Read et al. 2013), the current research programme was designed to incorporate perspectives from multiple system users, including car drivers, pedestrians, cyclists, motorcyclists and train drivers. It was also designed to capture road user behaviour across diverse types of rail level crossings, as there is no single design that represents a prototypical crossing (Edquist et al. 2009). Further, although there is a strong focus on novice driver safety in the broader road safety literature (Curry et al. 2011, Hatakka et al. 2002, Mayhew et al. 1998, McKnight and McKnight 2003), there was little research examining differences between novice and experienced drivers when specifically interacting with rail level crossings. Therefore, this comparison was considered within this phase of the research programme.

As a final point of note, the review of Edquist et al. (2009) highlighted the need for better data in this area, with part of the shortfall related to the methods used to study rail level crossing safety in previous research. The methods used in this research programme were selected as they provide a much greater level of detail than collected previously, particularly with respect to the cognitive processes underlying road users' behavioural choices. For example, the use of eye tracking in instrumented vehicles provides much richer data than conventional observational studies where the behaviour of the driver is inferred by observers recording at the roadside.

The methods used to better understand end user behaviour included the following:

- *On-road instrumented vehicle studies of driver behaviour*: Two on-road studies were completed: one focussing on urban active rail level crossings and the other focussing on active and passive crossings in a rural area. Participants were required to drive a pre-specified route that encompassed several rail level crossings, using an instrumented vehicle fitted with cameras and data logging equipment. Eye and head movements were recorded to measure drivers' allocation of visual attention, and their situation awareness was measured through provision of concurrent verbal protocols (i.e. thinking aloud during the drive).
- *Cognitive task analysis interviews of driver behaviour*: The on-road studies also involved a post-drive cognitive task analysis interview using the Critical Decision Method (CDM; Klein et al. 1989). CDM uses a series of structured prompts to facilitate recall of past events and probe factors that shaped decision-making. The interviews focussed on participants' decision-making at rail level crossings (i.e. whether to proceed, slow or stop).
- *Diary study of road user behaviour*: A diary study was used to capture data from multiple road user groups across diverse geographic locations. Drivers, pedestrians, cyclists and motorcyclists recorded their daily interactions at rail level crossings over a 2-week period. If participants encountered a train and/or active warnings, they were asked to recount this experience in detail. Diary questions were adapted from CDM probes to capture the factors that influenced decision-making.
- *Subject-matter expert interviews and in-cab familiarisation*: Discussions were held with two train drivers and one rail subject-matter expert to gather information regarding train driver behaviour at rail level crossings and their perceptions of road users' behaviour at rail level crossings. Researchers also participated in train cab rides through urban and rural areas to gain familiarisation with the train driving task and better understand the train driver perspective.

- *Observations of pedestrian and cyclist behaviour*: Structured on-site observations of pedestrian and cyclist behaviour were undertaken at seven urban rail level crossing sites (see Box 2.1). The aim was to understand the range of behaviours exhibited by users of the pedestrian infrastructure at rail level crossings and identify potential factors influencing behaviour. Due to space constraints, we do not report the findings in this chapter, but further information about the observations can be found in the work of Read et al. (2014).

The remainder of this chapter summarises the methods, results and conclusions from the primary data collection activities.

4.2 ON-ROAD STUDIES

A typical on-road instrumented vehicle study involves having participants drive a pre-specified route in a vehicle that is fitted with equipment that records vehicle parameters (e.g. driver inputs, vehicle speed, braking) as well as video footage of the road environment and the driver's behaviour. The route is specifically planned to include road features of interest, for the current project that meant designing routes that included multiple rail level crossings (Lenné et al. 2013). Similar methods have been used to study a range of research questions, including human error and driver distraction (Young et al. 2013a, b), situation awareness conflicts between different road users (Salmon et al. 2013d, 2014a, b) and the impact of road design on road user behaviour (Cornelissen et al. 2015, Salmon et al. 2014b, Young et al. 2013).

On-road studies using instrumented vehicles are advantageous because they include multiple measures of driver behaviour, which provide both an objective record of how individuals behave and their subjective interpretation of the situation and their self-reported reasons for behaviour. Objective measures of behaviour (e.g. speed, braking, lateral position) are provided by the vehicle data logger, as well as through video cameras that record the situational traffic conditions and the driver's behaviour. Eye and head movements can also be analysed to reveal drivers' allocation of visual attention, that is, where and when they looked for trains on approach to a rail level crossing. Subjective interpretation of the objective behaviour can be obtained by asking participants to provide concurrent 'think-aloud' verbal protocols during the drive, to provide insight into their situation awareness.

The on-road studies were designed to collect data across a range of different contexts, to better understand factors that influence drivers' behaviour and decision-making at rail level crossings. To facilitate this, two studies were planned: one in an urban area and the other in a rural region. The need to incorporate multiple rail level crossings strongly influenced the selection of data collection sites, as both studies needed to occur in locations with a high density of rail level crossings. The urban area selected was the south eastern suburbs of Melbourne (population 4.5 million), the state capital of Victoria, Australia, which has many actively controlled rail level crossings, all equipped with half boom barriers, flashing lights and bells. The rural area selected was the city of Bendigo (population 90,000), approximately 150 km north-west of Melbourne. Bendigo was chosen as three long-distance train lines intersect in the city centre, so it was possible to design an on-road test route that began in central Bendigo, with boom-controlled rail level crossings, and then quickly proceed into suburban and rural areas where crossings have passive controls or flashing lights only.

4.2.1 PARTICIPANTS

Twenty drivers (12 novice, 8 experienced) completed the urban study and 22 drivers (11 novice, 11 experienced) completed the rural study. Novice drivers were aged 18–22 years and held a provisional driver's licence. Experienced drivers were aged 29–55 years and held a full, unrestricted driver's licence. Each group contained approximately equal numbers of males and females.

4.2.2 INSTRUMENTED VEHICLE

Participants drove Monash University Accident Research Centre's On-Road Test Vehicle (ORTeV), an instrumented vehicle that combines a 32-channel Controller Area Network (CAN) interface, global positioning system (GPS) logger and seven unobtrusive cameras that record the driver, vehicle cockpit and forward, side and rear views external to the vehicle. Data collected included GPS location, vehicle heading, travel speed and brake pressure. In the rural study, a head-mounted eye tracking system (an Arrington Binocular Scene Camera 60 Hz) was used to record visual scanning behaviour.

4.2.3 TEST ROUTES

The urban test route was 11 km (20–25 minutes) and encompassed six rail level crossings, all with boom barriers, flashing lights and bells (see Figure 4.1). The route comprised arterial roads (80, 70 and 60 km/h) and shopping strip (40–50 km/h) areas.

The rural test route was 31 km (40 minutes) and encompassed nine rail level crossings: five with boom barriers, flashing lights and bells; one with flashing lights and bells only; two with Stop signs only; and one with a Give Way sign only (see Figure 4.2). Participants drove through one Stop crossing twice (once from each direction), for a total of 10 rail level crossing encounters. The route comprised varying road types, including city streets, shared spaces, suburban residential streets, highways and unsealed roads, with speed limits ranging from 40 to 100 km/h.

All drives were completed on weekdays at off-peak times (10:00 am and 2:00 pm for urban; 10:00 am and 1:30 pm for rural), to ensure participants experienced similar traffic conditions.

4.2.4 DATA COLLECTION PROCEDURE

Participants were recruited in the local area of each study, to increase the likelihood that they would have some familiarity or experience with the crossings in the study route. Each participant completed a single 2.5-h testing session, which began with providing informed consent and completing a brief demographic questionnaire.

FIGURE 4.1 Map of the urban on-road study route, which encompassed 11 km of roadway in the south eastern suburbs of Melbourne, Australia. Encircled numbers represent the locations of rail level crossings. All crossings were protected with boom barriers, flashing lights and bells.

The participant then received training in the verbal protocol procedure, which included demonstration, practice and feedback regarding how to provide appropriate verbal protocols (Young et al. 2015; see the appendix for example instructions). The aim of this training was to ensure participants verbalised cognitive tasks they were undertaking (e.g. 'I'm checking that the light is still green') rather than physical processes (e.g. 'I turned on the indicator').

Participants then completed the test route while concurrently providing their verbal protocol. In the urban study, participants drove alone with a route map, which

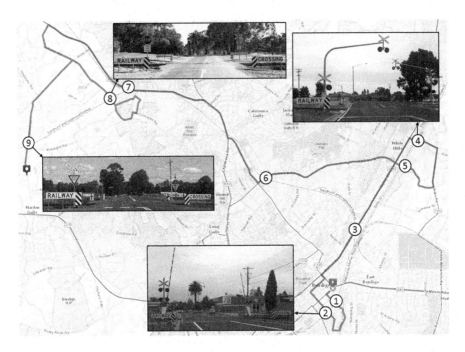

FIGURE 4.2 Map of the rural on-road study route, which encompassed 31 km of roadway beginning in Bendigo, Australia, and extending to surrounding rural areas. Encircled numbers represent the locations of rail level crossings. Five crossings were protected with boom barriers, flashing lights and bells (crossings 1, 2, 3, 6 and 7), one was protected with flashing lights and bells (crossing 4), two were protected with Stop signs (crossings 5 and 8) and one was protected with a Give Way sign (crossing 9). Participants encountered crossing 8 twice, once from each direction.

they had memorised beforehand. Two researchers followed at a distance in a separate vehicle, to assist if needed. In the rural study, participants were accompanied by two researchers in the vehicle, with the front seat observer providing navigation instructions.

4.2.5 Data Sources

The data collection methodology provided numerous data sources, including vehicle parameters, verbal protocols, eye glances and head movements. As the route characteristics and traffic conditions varied markedly between urban and rural environments, different approaches were adopted for data analysis across the two studies:

- Urban study analyses focussed on comparing train-present and train-absent conditions, as all rail level crossings had similar infrastructure and environments. Heavy surrounding traffic and restricted sight distances of approaching trains precluded meaningful analysis of speed profiles and head checks on approach to rail level crossings.

- Rural study analyses focussed on comparisons between different rail level crossing infrastructure features (i.e. active vs. passive) in train-absent conditions, as most drivers never encountered a train. Levels of surrounding traffic were low, so speed and stopping behaviour were relatively unconstrained.

Verbal protocols from both studies were analysed using Leximancer™ content analysis software to generate semantic networks representing participants' situation awareness on approach to rail level crossings (Salmon et al. 2013a). The semantic networks generated through this analysis process revealed the most prominent concepts that participants mentioned on approach to rail level crossings and the connections between them. An example network is shown in Figure 4.3.

4.2.6 KEY FINDINGS

Across both studies, there were 340 rail level crossing encounters: 120 at urban crossings, 132 at actively-controlled rural crossings and 88 at passively controlled rural crossings. Trains were present for 30% of urban encounters and 1% of rural encounters. No participants encountered trains at passive rail level crossings.

4.2.6.1 Urban Rail Level Crossings

Verbal protocol analyses revealed the most prominent concepts in drivers' situation awareness networks on approach to rail level crossings (Young et al. 2015). These were compared between train-present and train-absent encounters, and between experienced and novice drivers, to reveal the areas of commonality and divergence. Less than one-third of concepts were shared between all networks. Only two of

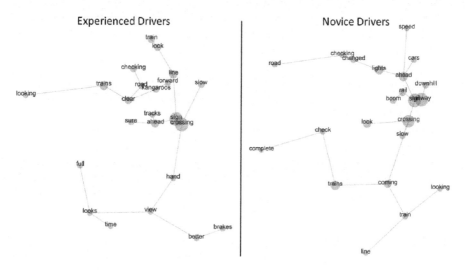

FIGURE 4.3 Situation awareness networks for experienced and novice drivers on approach to rural Stop-controlled rail level crossings. Shaded circles represent distinct concepts, and lines represent connections between concepts.

these common concepts were related to the rail level crossing (i.e. 'train', 'crossing'), with most shared concepts relating to the surrounding traffic environment (e.g. 'cars', 'road', 'lane', 'lights', 'front', 'left').

Examination of unique concepts revealed some notable differences as a function of driving experience and train presence. Experienced drivers attended to more diverse aspects of the environment, particularly non-car road users (e.g. pedestrians, trams), regardless of whether a train was approaching. Novice drivers demonstrated less attention towards non-car road users, and only mentioned the concepts related to monitoring of pedestrian traffic (e.g. 'pedestrians', 'people', 'door') when there was no train approaching. This suggests that the additional cognitive load of dealing with a train restricted novice drivers' ability to monitor other road users.

Analyses of visual scanning behaviour revealed that drivers spent a great majority of their time (87%–91%) looking at the forward roadway when approaching rail level crossings, regardless of train presence or driving experience. Experienced drivers spent a greater proportion of time glancing at the rear-view mirror and footpath/parking areas left and right of the road, whereas novice drivers spent more time looking at the speedometer and in-vehicle areas. These findings are consistent with the broader literature on eye movements in driving, which has demonstrated that drivers focus predominantly on the forward roadway, and that experienced drivers show more extensive horizontal scanning than novices (Chapman and Underwood 1998, Falkmer and Gregersen 2005, Underwood et al. 2002, Underwood 2007). It is also consistent with the verbal protocol analyses, suggesting that experienced drivers distribute their situation awareness more broadly, whereas novice drivers focus on their vehicle and forward vehicular traffic.

4.2.6.2 Rural Rail Level Crossings

Rural rail level crossing environments were much less visually cluttered than those in urban areas and therefore the analyses of rural data focussed primarily on how drivers searched for information (e.g. head checks made to confirm the presence or absence of a train) and their situation awareness, as indicated by their verbal protocols. Data were compared between novice and experienced drivers, across four types of crossings: Give Way sign; Stop sign; Flashing lights and bells; and Boom barriers, flashing lights and bells.

Head check data revealed stark differences in information seeking behaviour between passive and active crossings. Drivers made more head checks and spent longer visually searching for trains, at passive crossings compared with active crossings (see Figure 4.4a and b for the number and duration of head checks, respectively). There were no differences in the number and duration of head checks when comparing the two types of passive crossings (i.e. Give Way vs. Stop sign) or the two types of active crossings (i.e. Flashing lights vs. Boom barriers). However, the *timing* of head checks differed between Give Way and Stop sign crossings, with drivers beginning and ending the head check process much earlier on approach to Give Way rail level crossings (see Figure 4.4c). This reflects the fact that, in Australia, Stop signs are placed at rail level crossings with restricted sight distance (Standards Australia 2007).

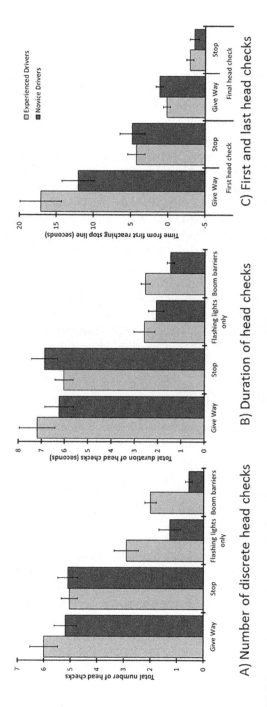

FIGURE 4.4 Patterns of head check behaviour by driver experience and rail level crossing infrastructure: (A) number of discrete head checks; (B) duration of head checks; (C) first and last head checks. Participants made more head checks (A) and spent longer checking (B) at passive rail level crossings (Give Way and Stop signs) compared with active rail level crossings (boom barriers or flashing lights only). At active rail level crossings, novice drivers made fewer head checks than experienced drivers. As shown in (C), head checks occurred later (i.e. shorter distance from the crossing) for Stop compared with Give Way rail level crossings.

Patterns of head checks at passive crossings did not differ between novice and experienced drivers. However, at actively controlled crossings, novice drivers spent less time checking for trains and made fewer head checks than experienced drivers (see Figure 4.4a and b). In fact, four novice drivers (36%) made no head checks on approach to any of the active rail level crossings, and only three (27%) made head checks at all active crossings, whereas most experienced drivers checked at all level crossings, regardless of whether active or passive infrastructure was present.

These findings imply that novice drivers greatly rely on active infrastructure to guide their decision-making at rail level crossings, which could have catastrophic consequences if signals fail or are unavailable. This was confirmed through verbal protocol analyses revealing differences between experienced and novice drivers' situation awareness on approach to rail level crossings with active and passive infrastructure. Eight networks were generated, representing experienced and novice drivers' situation awareness on approach to each of the four crossing types (i.e. Boom barriers, Flashing lights, Stop sign and Give Way sign). These networks revealed fundamental differences between novice and experienced drivers, especially when approaching passive crossings. Experienced drivers emphasised the need to make their own visual checks, regardless of whether the rail level crossing included active infrastructure, whereas novice drivers expected active signals would be present at crossings to warn them of approaching trains. An example of this is shown in Figure 4.3, which shows the Stop sign situation awareness networks. The novice driver network includes several concepts relating to active infrastructure (i.e. 'lights', 'boom'), which were not present at the Stop crossing, whereas these concepts do not appear in the experienced driver network.

4.3 COGNITIVE TASK ANALYSIS INTERVIEWS

Instrumented vehicle studies provide contemporaneous measurement of participants' objective behaviour and situation awareness during a drive. This provides insight into their actions, but additional methods are needed to fully explain why they chose that course of action. To delve deeper into road users' decision-making, researchers can conduct retrospective cognitive task analysis interviews that probe the key aspects of the decision-making process. Cognitive task analysis interviews can be used in conjunction with other methods, such as in the current project where on-road study participants completed cognitive task analysis interviews after their drive, or as a stand-alone research tool. In some circumstances, stand-alone interviews are the only feasible approach for data collection. However, respondents may provide idealised or non-specific answers that cannot be verified without an objective record of the events in question (Klein 1993). For this reason, the optimal approach is to use a combination of data collection methods.

There are several approaches to cognitive task analysis, but for studies concerned with naturalistic decision-making processes, a leading method is CDM (Klein et al. 1989, Klein and Armstrong 2005), which uses a series of structured prompts to aid recall of past events and explore factors that shaped decision-making. The CDM prompts explore how people assess the situation, determine that a decision must be made and formulate appropriate actions (Klein 1993). This process highlights the most critical information that users employ to make judgements and decisions, which should ideally be readily accessible through the system's design. Thus, a major advantage of studying naturalistic decision-making is that it can help designers to create systems that better reflect users' information processing needs (O'Hare et al. 2000).

CDM has been used to describe and assess naturalistic decision-making across varied domains, including aviation (Plant and Stanton 2013), military operations (Rafferty et al. 2012) and health care (Galanter and Patel 2005). Notably, these domains all involve highly trained personnel in safety-critical situations. Road transport differs from the traditional CDM context in that road users are not necessarily 'experts' at using a given transport mode. However, road users often face situations that resemble classic naturalistic decision-making paradigms, especially at rail level crossings: they must make the decision to stop or proceed under time pressure and dynamic conditions, with incomplete information, ill-defined goals and often poorly defined procedures. For this reason, CDM-based approaches have been gaining popularity as a method for understanding road user decision-making (Young et al. 2015). Further, CDM was chosen given the interest in novice-experienced driver differences and the role of expectancy in decision-making at rail level crossings, issues that we explored through applying Klein's (1993) recognition-primed decision (RPD) model.

4.3.1 INTERVIEW PROCEDURE

Interviews were conducted immediately after completing the on-road test routes described in Section 4.2. Each participant completed two interviews, with each interview focussed on their decision-making (i.e. whether to stop, slow, or proceed) at a single rail level crossing. In the rural study, each participant completed one interview about a passive crossing (i.e. Give Way or Stop sign only) and another interview about an active crossing (i.e. Boom barriers and/or Flashing lights). In the urban study, each participant completed one interview about a rail level crossing at which they encountered train or activated warnings, and another interview about a crossing at which they did not encounter a train or activated warnings. If participants encountered no trains during the urban route, or encountered trains at all rail level crossings, only one interview was conducted.

Individual interviews took 15–32 minutes for rural rail level crossings and 8–22 minutes for urban rail level crossings. This is considerably shorter than the typical CDM duration of 45–90 minutes cited for other domains (Klein 1993, Wong 2004), which reflects the fact that the rail level crossing encounters were typically short duration events (<1 minute if no train was present) and usually straightforward.

CDM is a flexible technique in that specific prompts can be selected to address attributes of interest. For this study, prompts were selected to provide a comprehensive description of the rail level crossing encounter, including the participant's goals, how they assessed the situation, the information they used to make their decision and their mental representations of other alternatives (see Table 4.1 for a full list).

TABLE 4.1
List of CDM Interview Prompts Used

Prompt	Interview Question
Incident description[a]	Describe your encounter in as much detail as possible below, including what you did, why you did it, the traffic conditions, how you knew a train was coming and how the encounter unfolded? Try and include the information you were using and any things that influenced what you did (e.g. time pressure, other road users, experience of this crossing)
Goal specification[a]	What were your specific goals at this point in time?
Assessment[a]	Suppose you were to describe the situation at this point to someone else. How would you summarise the situation?
Cue identification[a]	What features were you looking for when you formulated your decision?
	How did you know that you needed to make the decision? How did you know when to make the decision?
Expectancy	Were you expecting to make this sort of decision during the course of the event? Describe how this affected your decision-making process.
Options	What courses of action were available to you? Were there any other alternatives available to you other than the decision you made?
	How/why was the chosen option selected? Why were the other options rejected?
	Was there a rule that you were following at this point?
Influencing factors[a]	What factors influenced your decision-making at this point?
	What was the most influential factor that influenced your decision-making at this point?
Situation awareness[a]	What information did you have available to you at the time of the decision?
Situation assessment[a]	Did you use all of the information available to you when formulating the decision?
	Was there any additional information that you might have used to assist in the formulation of the decision?
Information integration	What was the most important piece of information that you used to formulate the decision?
Experience[a]	What specific training or experience was necessary or helpful in making this decision?
Mental models[a]	Did you imagine the possible consequences of this action?
	Did you create some sort of picture in your head? Did you imagine the events and how they would unfold?
Decision-making[a]	How much time pressure was involved in making the decision?
	How long did it actually take to make this decision?
	Was there any stage during the decision-making process in which you found it difficult to process and integrate the information available?
Conceptual model[a]	Are there any situations in which your decision would have turned out differently?
Basis of choice	Do you think that you could develop a rule, based on your experience, which could assist another person to make the same decision successfully?
Analogy/generalisation	Were you at any time reminded of previous experiences in which a similar/different decision was made?

[a] Prompts that were also used in the diary study (see Section 4.4).

4.3.2 DATA ANALYSIS

Audio recordings of interviews were collected using a digital Dictaphone and recordings were transcribed verbatim, so that the interview transcripts could be used for analysis. For this current study, responses were coded manually using Klein's (1993) RPD model as a framework.

The RPD model was developed after in-depth studies with fire-ground commanders revealed that they typically select a course of action immediately after a rapid assessment of the situation, without comparing or evaluating options (Klein 1993). In its simplest form, RPD involves a 'simple match' where the decision-maker recognises the situation, selects a previously successful action and implements it. Recognition is based on four aspects: the individual's goals, their expectancies, relevant cues in the environment, and possible actions. If the situational context is changing or unfamiliar, or expectancies are violated, recognition becomes more complex (see Figure 4.5). These complexities cause the individual to seek further information, reassess and mentally simulate a possible action to determine its likelihood of succeeding before implementation.

4.3.3 KEY FINDINGS

The CDM interviews provided a wealth of data, which enabled exploration of differences in decision-making as a function of driver experience (novice or experienced), driving environment (urban or rural), rail level crossing infrastructure (active or passive) and train presence. This revealed key differences in decision-making as a function of driving experience and the rail level crossing infrastructure and environment.

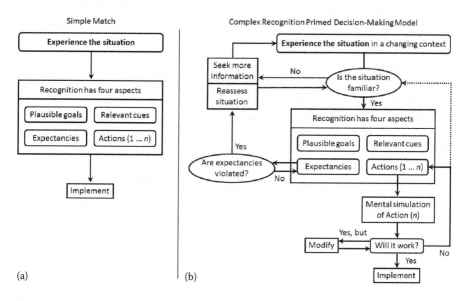

FIGURE 4.5 Schematic of Klein's (1993) RPD model, showing the cognitive processes for a simple match (a) and more complex changing contexts (b).

4.3.3.1 Novice versus Experienced Drivers

Comparisons between novice and experienced drivers revealed some fundamental differences in their expectancies and use of relevant cues, especially at passive rail level crossings. These differences reflect underlying experiences; all participants in the experienced driver group had prior experience in driving across passive crossings, albeit infrequently, whereas half of the novice group had never encountered a passive crossing before. This meant that experienced drivers could proceed in a straightforward manner through the RPD model where they recognised the situation, made a rapid assessment and then selected a course of action. This was especially true at the Stop sign crossings, where most drivers made a simple match decision to come to a complete or rolling stop and then visually check for trains before proceeding.

For novice drivers, passive crossings violated their expectancies that all rail level crossings would have active controls. This also meant that the primary informational cues they sought (i.e. flashing lights and boom barriers) were unavailable, prompting them to seek more information so they could fully assess the situation. This meant that they could not follow the same RPD path as more experienced drivers. Notably, novice drivers tended to mention that the absence of active controls forced them into a different information-seeking strategy, whereby they had to make their own checks for a train rather than relying on the rail level crossing infrastructure. In contrast, experienced drivers commented that they habitually made their own visual checks regardless of the rail level crossing infrastructure. Indeed, in several instances, experienced drivers in the rural study were unable to recall at the interview whether specific crossings had active or passive infrastructure.

Although differences between novice and experienced drivers were most pronounced at passive rail level crossings, differences were also observed at active rail level crossings, particularly in the more complex urban environment. For both novice and experienced drivers in the urban study, their goals and cues were focussed on the behaviour of other vehicles. When no train was present, drivers were concerned with maintaining a smooth traffic flow, whereas when a train was present, they used other vehicles as cues to slow and stop. However, novice drivers were focussed on maintaining an appropriate travel speed (i.e. slow enough to stop if required, but fast enough to maintain traffic flow), whereas experienced drivers were more concerned with ensuring they would not be stopped on the train tracks if traffic ahead stopped. Several experienced drivers reported using estimates of the space available on the far side of the crossing as a cue to proceed (i.e. it is safe when there is enough room to clear the tracks), whereas no novice drivers mentioned this factor as influencing their decision-making.

4.3.3.2 Urban versus Rural Environments

Even though drivers were significantly more likely to encounter trains at urban rail level crossings compared with those in rural locations, they were less likely to view urban crossings as posing a safety hazard. In rural settings, drivers' primary goals were to check for trains, avoid hitting a train, to cross the rail level crossing safely and to watch the lights. Their decision-making focussed first and foremost on negotiating the rail level crossing: 'obviously, the railway crossing was the biggest danger.

[My goal was] avoid getting hit by a train'. Participants, especially experienced drivers, reported using their own visual checks as cues to guide their decision-making, even for crossings with active infrastructure. This self-reliance reflected a common belief that signals are unreliable, and thus, inactive signals should not be interpreted as meaning the situation is safe.

In contrast, drivers in urban settings expressed an extremely diverse range of goals, including many that were seemingly unrelated to the rail level crossing, such as monitoring parked cars and pedestrians, maintaining the traffic flow and keeping to a safe speed. Several drivers directly stated that the presence of the rail level crossing did not influence them, even when a train was present: 'the level crossing itself didn't actually factor much into my decision-making'. The most common goals concerning the rail level crossing were to keep the tracks clear and traverse the crossing, with a minority aiming to cross before the lights changed or a train arrived. The primary safety concern influencing decision-making was avoiding queuing or short stacking, and keeping the tracks clear: 'my goals were to be really careful because I hate level crossings and I'm always really, really careful to make sure that I don't have any potential to get stuck in the level crossing'. The decision-making processes of urban drivers were also closely linked to the behaviour of other road users: other road users provided cues to alert them to the presence or absence of a train, and constrained their actions (e.g. by stopping and thus forcing others behind them to stop).

4.4   DIARY STUDY

Both the on-road studies and cognitive task analysis interviews provided detailed data about road users' behaviour, which is necessary to understand the factors that shape actions and decisions at rail level crossings. A distinct limitation of these methods is that they are usually restricted to relatively small sample sizes, due to the resources required to collect data. Further, they are typically skewed towards car drivers. An alternative is to adapt the interview prompts to a survey format, which allows researchers to collect data from a larger, more diverse participant group and to include multiple road user groups. Adaptation of CDM in a written format has been successfully used to examine drivers' situation awareness and decision-making previously (Walker et al. 2009a) and was adopted here to examine the differences in decision-making between different road user groups at rail level crossings (Beanland et al. 2013, 2016). Regular rail level crossing users were recruited for a prospective longitudinal study, in which they were required to document their experiences at level crossings in a daily diary over a 2-week period.

4.4.1   PARTICIPANTS

Participants were 50 car drivers, 39 motorcyclists, 42 cyclists and 35 pedestrians who used rail level crossings at least once per week. All resided in Victoria, Australia; most lived in the Melbourne metropolitan area, with a minority (38% of car drivers, 28% of motorcyclists, 6% of pedestrians and 2% of cyclists) residing in rural and regional areas.

4.4.2   Survey Format and Content

The survey was developed in two formats: a paper-based and an online version deliv-ered via SurveyGizmo™ (www.surveygizmo.com), which could be used on comput-ers and portable electronic devices (e.g. smartphones). The content was identical in both formats, but the online survey had more flexible presentation (i.e. presenting only relevant response options, based on answers to previous questions). Research directly comparing paper and online surveys with identical questions and compara-ble participant demographics has not found any difference in the responses provided (Burdett et al. 2016).

Participants completed an initial demographic questionnaire, followed by daily 'diary' surveys for 2 weeks. In the daily survey, they recorded the number and type of rail level crossings they crossed and how many times they encountered a train and/or acti-vated warning signals that day. For days where they encountered a train and/or activated warnings, respondents were prompted to describe this encounter (or one encounter, if they had multiple) in detail by following through a series of CDM prompts. The full daily survey took 10–15 minutes to complete. Participants using the online survey were sent e-mail reminders to reduce attrition rates.

The CDM prompts were adapted from those used in the cognitive task analysis interviews (see Table 4.1), although fewer prompts were used to reduce the time required to complete responses. A key difference between surveys and interviews is that for surveys, it is not possible to follow up on comments that the participant makes, so a broad range of prompts were used to elicit the required information. Survey respondents may also provide less information than interviewees, because it is easier to provide responses verbally than in writing. To overcome this limitation, the survey included pre-defined response options to ease the reporting burden on participants. The first two prompts (Incident description and Goal specification) and the final prompt (Conceptual model) used open-ended questions where participants were required to write a response, but for the other prompts, participants could select one or more options from a list, with provision to specify additional options where appropriate. Response options were generated based on previous research and con-sultation with system documentation such as standards (e.g. for Cue identification, the options included all possible infrastructure items and likely environmental ele-ments, such as other road users).

4.4.3   Key Findings

Participants' self-reported travel habits suggested that most were commuters, spend-ing on average 9.2 h per week using their nominated travel mode and encountering an average of one to two rail level crossings per day, with both trip frequency and rail level crossing exposure being significantly higher on weekdays compared with weekends. Car drivers encountered more rail level crossings than other road users, which likely reflects longer trips (i.e. a driver could cross multiple rail level crossings in a route, whereas a pedestrian might only cross one *en route* to the train station).

Most encounters with trains and/or active signals occurred within metropoli-tan Melbourne, with very few in rural areas. Responses to the Assessment prompt revealed that many road users experienced demanding conditions on approach to the rail level crossing, including the presence of heavy traffic and multiple other road users, being in a hurry, being the first road user to reach the crossing when the signals activated or having to stop or turn immediately before or after the rail level crossing. Most encounters reported (90%) involved compliant behaviours, which included slowing or stopping completely to wait for the train to pass, or re-routing to avoid waiting for the train, whereas 10% of encounters involved non-compliant behaviours. Under local laws, it is illegal to cross when the warning signals are activated or when a train is visible, and there would be a danger of a collision occurring, so both actions were classed as non-compliance.

4.4.3.1   Predictors of Non-Compliant Behaviour

Several factors predicted the likelihood of non-compliant behaviour, including the type of road user and the conditions during approach. Pedestrians were more likely than all other road users to report non-compliant actions: 19% of pedestrians reported crossing in front of a train, compared with 4% of other road users. Respondents were also more likely to commit violations if they were the first to arrive at the rail level crossing, were in a hurry and/or felt time pressured, but were less likely to violate if there was heavy surrounding traffic.

There were also significant differences between compliant and non-compliant road users with respect to the experience and influencing factors that informed the individual's decision-making. Road users who committed violations were more likely to base their decision-making on the knowledge of the specific crossing and their own acceleration or braking abilities (i.e. whether they could go fast enough to complete the crossing in time). In contrast, compliant road users were more likely to base their decision on the knowledge of road rules, general knowledge of rail level crossings and information derived from other vehicles and the crossing infrastructure (i.e. boom barriers, flashing lights and bells).

4.4.3.2   Differences between Road Users

In addition to differences between compliant and non-compliant road users, there were also differences between road user groups in terms of the most important influencing factors that shaped their decision-making (regardless of the decision made). The most striking discrepancies emerged when comparing motorists with non-motorists, and when comparing pedestrians with other road users.

When comparing motorists (i.e. car drivers and motorcyclists) with non-motorists (i.e. cyclists and pedestrians), it was apparent that motorists' decision-making was influenced more by flashing lights, the behaviour of other vehicles and (to a lesser extent) signs at the crossing. In contrast, non-motorised users were more likely to be influenced by hearing a train, which is probably attributable to the fact that it is physically easier for these users to hear audible cues in the environment (whereas vehicle cabs and motorcycle helmets are likely to mask these sounds).

Pedestrians also exhibited several additional differences to other road users. Specifically, compared with the other road user groups, pedestrians were less likely to be influenced by boom barriers and traffic lights, and were less likely to base their decision-making on the knowledge of road rules. However, pedestrians were more likely to base their decision-making on the knowledge of train speeds and were more likely to be influenced by the sight of a train and the operation of pedestrian gates.

4.5 INPUT FROM SUBJECT-MATTER EXPERTS

The primary data collection activities undertaken in Phase 1 of the research programme were the on-road studies, cognitive task analysis interviews and diary records, which were designed to comprehensively study how road users interact with rail level crossings. Because the legislation governing road user behaviour at rail level crossings requires the road user to give way to the train, road user compliance with warnings and signs is a key concern for research aiming to improve safety in this area. However, accidents at rail level crossings have extensive negative psychological impacts on train drivers. In Australia, train drivers and assistants were the occupational group with the highest number of mental stress claims for the years 2008–2009 and 2010–2011 (Safe Work Australia 2013), and studies have highlighted the impact of a 'person under the train' incident on train drivers through measures such as anxiety, depression, lack of meaning in life, loss of control and sense of safety and sense of guilt (Mehnert et al. 2012). Furthermore, a systems view emphasises the need to take account of all the interactions within a system, and it was considered important to understand the train driver behaviour to gain insights into potential redesign opportunities.

As such, interviews with subject-matter experts were employed as an additional data source to provide an understanding of rail level crossing functioning from the rail industry perspective. This included discussions with two train drivers and one human factors practitioner working in the rail industry, who had over 10 years' experience working in risk management across safety-critical industries. Discussions focussed on train drivers' behaviour at rail level crossings and their perceptions of road user behaviour at rail level crossings. Researchers also participated in train cab rides through urban and rural areas to gain familiarisation with the train-driving task and better understand the train driver perspective.

4.5.1 Key Findings

The subject-matter expert interviews and in-cab familiarisation rides revealed the key aspects of train driver behaviour at rail level crossings, including their actions in monitoring infrastructure, observing road users and using route familiarity and knowledge in the performance of tasks.

4.5.1.1 Monitoring Infrastructure

When approaching rail level crossings, train drivers reported responding to whistle boards by sounding their whistle, checking for indications that predictors are used by the warning system and maintaining their speed where this is the case (to assist in achieving a consistent 25 second warning time for road users), and checking that the warnings to road users are working. Ensuring the operation of warnings is achieved by checking for healthy-state lights (where present) or otherwise looking to the flashing light assemblies to determine whether the system is functioning correctly. There are two failure modes for rail level crossing warning devices: 'fail safe' and 'fail unsafe'. Fail safe situations occur where a failure results in the activation of the crossing's warning signals until the failure is resolved. Fail unsafe situations, considered to be dangerous failures, occur when signals do not activate as designed; for example, where an approaching train is not detected (Wullems 2011). Train drivers reported actively checking to ensure the system is not in the 'fail unsafe' state, even though in most cases it is physically impossible for them to intervene or change their actions at this point due to the poor braking capability of trains. An obvious indication of the fail unsafe state would be if road users were still crossing the tracks, or if booms remained in the upright position, when the train was within the sight of the rail level crossing.

4.5.1.2 Monitoring Road Users

In addition to observing the signals, train drivers reported observing the behaviour of road users to assess compliance. These comments align with previous research findings that train drivers attempt to predict what road users will do based on observing their behaviour on approach (e.g. Davey et al. 2005). As with monitoring for failure states, these observations would typically occur too late for meaningful intervention. However, this gives train drivers knowledge regarding typical road user behaviour, which could inform reporting of near misses.

Train drivers also held a perception that road users having high familiarity with specific crossings can foster complacency, which can in turn lead to reckless behaviour. For example, they highlighted a case of a road user who was killed crossing a train line that had been closed for an extended period, but had been recently reopened.

4.5.1.3 Route Knowledge

Finally, the train drivers noted relying heavily on their route knowledge, combined with operating rules and procedures, when making decisions at rail level crossings. Whereas the standard procedure is to use whistle boards as an indicator of when to sound the train whistle, in reality most train drivers have an extensive knowledge of the routes they drive and can prepare for sounding the whistle before they actually reach the whistle board. This, coupled with their observations of road user behaviour, can lead them to adapt their own behaviour to the local context, by sounding the whistle for extended periods at rail level crossings that they perceive to be especially dangerous.

Train drivers' perceptions of danger may be based on their own or other drivers' observations of incidents, near misses and road user violations at rail level crossings. Assessments may also be based on the knowledge of the rail level crossing infrastructure and layout, as train drivers may sound the whistle for longer if it is a passive rail level crossing, or if either train drivers or road users have limited sight distance on approach due to the approach angle or roadside occlusions such as foliage and embankments. Finally, train drivers may adapt their speed on approach to crossings where they feel additional warning time is required.

4.6 SUMMARY

This chapter has described the various data collection activities we used to better understand behaviour at rail level crossings. This phase of the research programme collected a range of novel data covering multiple user groups that are often not considered, including cyclists, motorcyclists, pedestrians and train drivers. In addition, the integration of objective data (e.g. vehicle measures, eye tracking) with subjective data (e.g. cognitive task analysis interviews, verbal protocol analysis) provided a level of detail and understanding not previously achieved in this domain.

The data collected provided some key insights into the functioning of existing level crossing systems, which are as follows:

- There are differences in the way novice and experienced drivers interact with rail level crossings, consistent with the wider road safety literature.
- There are important differences in how drivers behave in urban and rural settings, and at rail level crossings with different types of controls (e.g. active signals, Stop sign, Give Way sign).
- Experience and expectancy play a key role in user behaviour at rail level crossings.
- Different types of road users (e.g. drivers, motorcyclists, cyclists and pedestrians) use different forms of information to make decisions when deciding whether to 'stop or go' at a rail level crossing.
- Train drivers actively monitor crossing warnings and road user behaviour, but have few options for avoiding a collision due to limited braking capabilities of trains.

In addition to the insights provided, the data were chosen as the ideal inputs for the systems modelling approaches undertaken in Phase 2 of the research programme. These modelling activities are described in Chapter 5.

5 A Systems Analysis of Rail Level Crossings

With Contributions from:
Christine Mulvihill, Guy Walker and Miranda Cornelissen

5.1 INTRODUCTION

This chapter describes the findings from a systems analysis of the existing rail level crossing system in Victoria, Australia. The purpose of the overall analysis was to:

1. Develop an in-depth description of the functioning of rail level crossing systems in Victoria, Australia.
2. Identify the key issues that currently do, or potentially could, threaten safety at rail level crossings.
3. Generate insights to inform the design of new safer and more efficient rail level crossing environments.

The analysis involved applying the following:

- *Cognitive Work Analysis* (CWA; Vicente 1999): CWA was applied to produce an in-depth description of rail level crossing system functioning and the factors influencing it, and to identify key findings to input into design processes.
- *Hierarchical Task Analysis* (HTA; Annett 2004): HTA was employed to produce a goal-based description of drivers' interactions with passive and active rail level crossings. In addition, the description was used to inform a human error identification process.
- *Systematic Human Error Reduction and Prediction Approach* (SHERPA; Embrey 1986): SHERPA was applied to identify the errors that drivers could make at existing rail level crossings in Victoria, Australia, and to explore the extent to which existing crossing environments exhibit error tolerance (i.e. are designed to prevent human errors, or provide opportunities for error identification, recovery or mitigation). Finally, the SHERPA analysis provided a means to identify the initial design ideas for improving behaviour and safety at rail level crossings.

5.2　CWA OF RAIL LEVEL CROSSING SYSTEMS

As outlined in Chapter 2, CWA (Vicente 1999) is a systems analysis and design framework that has previously been used both to analyse sociotechnical systems and to inform system design or redesign activities (see Bisantz and Burns 2008, Stanton et al. 2017). The framework describes the various constraints that influence behaviour within systems, with design applications often focussing on making system constraints visible to operators in a way that supports them to adapt and respond to unexpected situations and system disturbances. Another notable feature of the framework is that it is formative in nature, with its methods able to describe both how behaviour currently occurs (descriptive analysis) and how it could occur given the systems constraints, or with modification to these constraints (formative analysis). This formative component is especially useful for informing system design or redesign, as it enables consideration of different ways in which functions and affordances can be achieved.

The CWA framework comprises five analysis phases (Vicente 1999). In this application, only the first four phases (Work Domain Analysis [WDA], Control Task Analysis [ConTA], Strategies Analysis, Social Organisation and Cooperation Analysis [SOCA]) were employed. The fifth phase, Worker Competencies Analysis, was not applied as it was determined that the outputs from the first four phases provided a sufficiently in-depth account of rail level crossing systems to identify the areas of sub-optimal functioning and provide insights to inform redesign. Full details of the analysis are available in the work of Salmon et al. (2016b); a summary is presented in this chapter.

5.2.1　Analysis Approach

The aim of the CWA was to produce an in-depth description of rail level crossing system behaviour and the factors influencing it, and to generate design insights through formative analyses of rail level crossing systems. The data inputs and outputs from each phase are presented in Table 5.1.

Multiple analysts with significant experience applying CWA in varying domains (including defence, road transport, rail transport, aviation and maritime) were involved. The analysis was primarily conducted during a series of workshops where the analysts worked together to conduct each analysis phase. This collaborative process enabled any disagreements to be resolved *in situ* and ensured that multiple perspectives were offered. Of the four phases applied, the Strategies Analysis was the only phase that was undertaken by a single analyst; however, the remainder of the research team reviewed the analysis and its outputs.

The data used to inform the CWA were gathered during the data collection activities described in Chapter 4, along with additional activities such as documentation analysis (see Table 5.1 for an overview).

5.2.2　Work Domain Analysis

WDA, the first phase of CWA, was used to provide an event- and actor-independent description of existing active and passive rail level crossing systems in Victoria,

TABLE 5.1

CWA Phases, Outputs and Rail Level Crossing Examples

CWA Phase	Data Inputs	Outputs	Rail Level Crossing Example
Work Domain Analysis	• Document review and analysis (e.g. design standards, policy documentation) • On-road studies, including cognitive task analysis interviews • Diary study of road user behaviour • Subject-matter expert interviews • In-cab familiarisation • Observations of pedestrian behaviour • Subject-matter expert workshops to refine and validate initial models	Abstraction hierarchy model of the system, including functional purpose, values and priority measures, generalised functions and physical objects and their affordances	Work Domain Analysis model showing functional purposes of rail level crossing systems (e.g. provide access over rail line), values and priority measures (e.g. minimise collisions), functions (e.g. alert road user to presence of train) and physical objects (e.g. flashing lights) and their affordances (e.g. provide warning of train)
Control Task Analysis	• Diary study of road user behaviour • On-road studies, including cognitive task analysis interviews • Work Domain Analysis	Decision ladders showing decision-making process for different key decisions along with shortcuts made by experts Contextual activity template showing the functions that occur across different situations	Decision ladder showing information, goals and options related to the 'stop or go' decision at rail level crossings Contextual activity template showing which functions occur in different situations (e.g. road user at crossing, train at crossing) and which functions could occur through redesign efforts

(Continued)

TABLE 5.1 (*Continued*)
CWA Phases, Outputs and Rail Level Crossing Examples

CWA Phase	Data Inputs	Outputs	Rail Level Crossing Example
Strategies Analysis	• Diary study of road user behaviour • On-road studies, including cognitive task analysis interviews • Observations of pedestrian behaviour • Work Domain Analysis • Control Task Analysis	Strategies Analysis diagram depicting the different strategies that can be used to undertake control tasks	Strategies analysis diagram showing all the different ways in which different users (e.g. drivers, pedestrians, motorcyclists and cyclists) can come to a 'stop or go' decision at the rail level crossing and execute the task of stopping or proceeding
Social Organisation and Cooperation Analysis	• Diary study of road user behaviour • On-road studies, including cognitive task analysis interviews • Work Domain Analysis • Control Task Analysis • Strategies Analysis	Abstraction hierarchy, decision ladders and contextual activity templates shaded to show allocation of functions across different actors (human and non-human)	Abstraction hierarchy showing which different actors currently perform the different functions required (e.g. which human and non-human actors perform the function 'alert road user to presence of train')

Australia. The aim was to describe the purposes of rail level crossing systems and the constraints imposed on the behaviour of actors operating within them (Vicente 1999). This was achieved by constructing two abstraction hierarchies, which described active and passive rail level crossings at the following five conceptual levels:

1. *Functional purpose*: The overall purposes of the system and the external constraints imposed on its operation.
2. *Values and priority measures*: The criteria that organisations use for measuring progress towards the functional purposes.
3. *Generalised functions*: The general functions of the system that are necessary for achieving the functional purposes.
4. *Physical functions*: The functional capabilities and limitations of the physical objects within the system that enable the generalised functions.
5. *Physical objects*: The physical objects within the system that are used to undertake the generalised functions.

In the abstraction hierarchy, the lines connecting nodes at adjacent levels are known as 'means-ends links'. These links can be read using the 'why–what–how' relationship. Taking any node within the hierarchy as the 'what', nodes linked in the level above the node indicate why it is necessary within the system and nodes linked in the level below represent how the node is achieved (Vicente 1999). By tracing the means-ends links from the bottom of the hierarchy upwards, it is possible to analyse how individual components can have an impact on the overall purpose(s) of the system.

As the abstraction hierarchy diagrams are large and complex, a summary of the active rail level crossing WDA is presented in Figure 5.1. The passive rail level crossing abstraction hierarchy had the same functional purposes, values and priority measures, and generalised functions. Differences were found at the lower two levels, as passive crossings have no active warning devices (boom gates, flashing lights, bells) and thus do not possess the related physical functionalities around alerting the road user to the presence of a train.

As shown in the top level of Figure 5.1, six distinct functional purposes were identified. These are largely unsurprising; however, an interesting feature is the potential conflicts between them. For example, maintaining priority access for rail traffic while minimising delays to the road network represents incongruent functional purposes that are difficult to achieve simultaneously. This is especially the case at busy urban crossings where warnings may remain active for extended periods of time as multiple trains traverse a crossing. Delays for road users and pedestrians are one factor influencing non-compliant behaviours (i.e. driving around boom gates or crossing in front of train). Moreover, this conflict is likely to escalate in future given increasing road and rail traffic levels.

The 'values and priority measures' level of the abstraction hierarchy outlines the criteria that can be used to assess the system's progress towards achieving its functional purposes. Seven core values and priority measures were identified, including minimising collisions, injuries, fatalities, risk and road rule violations; maximising

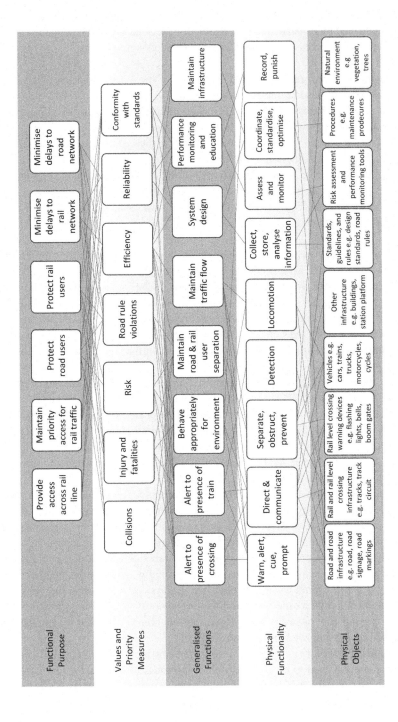

FIGURE 5.1 Summary of active rail level crossing abstraction hierarchy. (Adapted from Salmon et al., 2016b.)

efficiency and reliability of the crossing and achieving conformity with design standards. Although these values and priorities appear appropriate, it is questionable whether existing rail level crossings systems possess the systems required to assess the extent to which they are being met. For example, arguably the measures around road user non-compliance and near miss incidents are currently not well understood as there are few mechanisms for collecting and analysing this information. Although incidents are reported by train drivers, road users do not have dedicated mechanisms to report near misses and instances of non-compliance. Similarly, although there is a risk assessment tool and process for determining priority of upgrades from passive to active warnings in Victoria, both have attracted criticism in recent times (Coroner Hendtlass 2013, Salmon et al. 2013c). A coronial inquiry that investigated 26 deaths at rail level crossings in Victoria (including the Kerang accident described in Chapter 1) recommended that agencies '… improve the accuracy, content and relevance of data used in predictive risk analysis used to inform decisions about upgrading of level crossings …' (Coroner Hendtlass 2013, p. 136). It is therefore debateable whether accurate information is currently available to assess the level of risk associated with different crossings. An important implication from this level of the abstraction hierarchy is that road and rail organisations may not fully understand the extent to which rail level crossings are meeting key values and priorities.

The 'generalised functions' level of the abstraction hierarchy outlines the functions that must be achieved for rail level crossings to fulfil their functional purposes. This level emphasised the current design's focus on the requirement for road users to stop for trains. Several functions relate specifically to road users (i.e. alerting them to the rail level crossing and the presence of a train, ensuring that they behave appropriately), whereas other functions are associated with the separation of road and rail users, maintaining traffic flow and designing, monitoring and maintaining the rail level crossing environment. A key insight from this level is that various combinations of failed functions can lead to collisions. For example, the system failing to alert the road user to the presence of a train represents one failed function that can directly cause a collision; however, multiple functions undertaken inadequately could also interact to create a collision. For example, poor maintenance of the rail level crossing could lead to diminished warnings (e.g. faded passive signage and road markings on approach to the crossing, or boom arms failing to lower due to mechanical faults), which could lead to road users not being effectively warned about the presence of a train and not behaving appropriately.

A second key insight from this level is the profound influence that functions across the wider rail level crossing system have on behaviour and safety. For example, the analysis shows how functions such as 'system design' and 'performance monitoring and education' are required to achieve the system's overall functional purposes. This demonstrates how factors away from the level crossing contribute to collisions and, moreover, that design modifications need not focus exclusively on the crossing environment.

Finally, poorly supported functions can be identified. For example, performance monitoring has few means-ends links from the level below and, although maintenance of traffic flow has several links, often these are not well achieved in busy urban environments where there is a high frequency of train traffic and warnings are activated often. In addition, although warning devices such as flashing lights provide warnings to alert road users of approaching trains, these warnings do not always alert road users to the presence of a train even when they are operating as designed (e.g. as occurred in the Kerang collision described in Chapter 1).

The lower two levels of the WDA show the physical objects that the system comprises, along with their affordances. Physical objects were grouped into the following categories: road and road infrastructure (e.g. the road, kerb, lane markings), rail level crossing infrastructure (e.g. tracks, whistle board, train detection systems), warning devices (e.g. flashing lights, early warning signage), vehicles (e.g. cars, trucks, trains), other infrastructure (e.g. buildings), standards, guidelines and rules (e.g. road rules, road and design standards), risk assessment tools, procedures and the natural environment (e.g. vegetation, weather conditions).

5.2.3 CONTROL TASK ANALYSIS

The second CWA phase, ConTA, is used to examine the specific tasks that are undertaken to achieve the purposes, priorities and functions of a work domain (Naikar et al. 2006). ConTA-phase methodologies include the contextual activity template (CAT; Naikar et al. 2006) and the decision ladder (Rasmussen 1976, cited in Vicente 1999), both of which were used in the current project. The CAT is used to map control tasks across different situations, and to identify situations in which they are currently or could potentially be undertaken. The decision ladder is used to describe decision-making processes that can be adopted during different tasks, along with decision-making 'shortcuts' that can be made by experienced users.

5.2.3.1 Contextual Activity Template

The CAT (Naikar et al. 2006) was used to examine where control tasks are currently achieved at rail level crossings and where they could potentially occur. CATs were developed for active and passive rail level crossings; an extract of the active rail level crossing CAT is presented in Figure 5.2.

The first step in developing each CAT involved identifying the situations and control tasks within the rail level crossing system and determining how these should best be represented. The situations, shown along the horizontal axis in Figure 5.2, depict the spatially and temporally distinct approach phases that the road and rail users progress through as both approach the rail level crossing.

- For the rail user, the phases were defined as follows: pre-whistle board, at whistle board, at train detection device, at station pre-rail level crossing, traversing rail level crossing and post-rail level crossing.
- For the road user, the phases were defined as follows: pre-approach, on-approach, pre-boom gates, at boom gates/boom gates closing and on rail level crossing.

Combining these phases for the road and rail users led to a total of 66 situations to be analysed via the CAT.

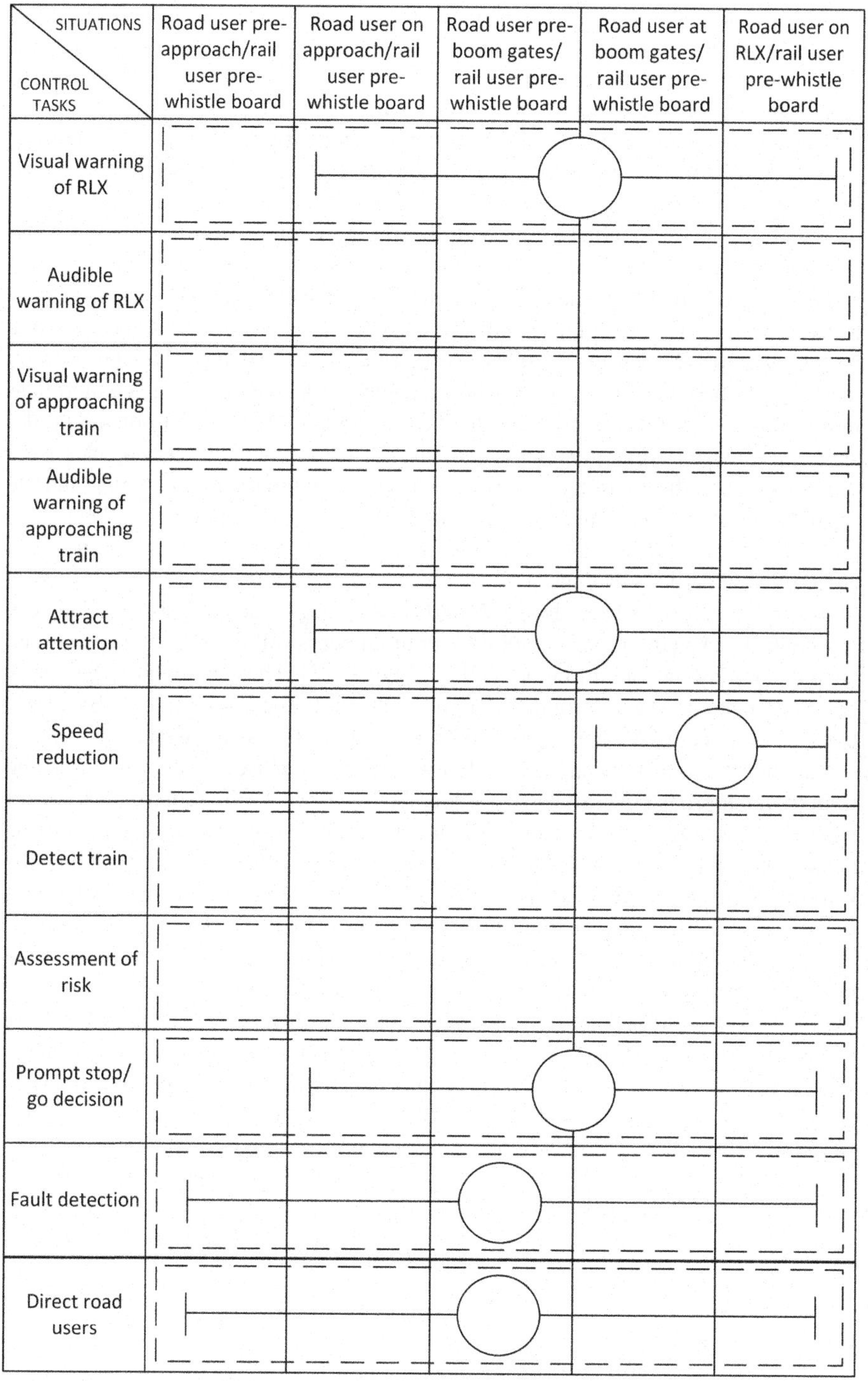

CONTROL TASKS \ SITUATIONS	Road user pre-approach/rail user pre-whistle board	Road user on approach/rail user pre-whistle board	Road user pre-boom gates/rail user pre-whistle board	Road user at boom gates/rail user pre-whistle board	Road user on RLX/rail user pre-whistle board
Visual warning of RLX		⊢	──○──	──┤	
Audible warning of RLX					
Visual warning of approaching train					
Audible warning of approaching train					
Attract attention		⊢	──○──	──┤	
Speed reduction				⊢──○	──┤
Detect train					
Assessment of risk					
Prompt stop/go decision		⊢	──○──	──┤	
Fault detection	⊢		─○──		──┤
Direct road users	⊢		─○──		──┤

FIGURE 5.2 Extract of CAT for active rail level crossings.

The control tasks for the CAT were taken from the 'physical functionality' level of the abstraction hierarchy. These are shown along the vertical axis of Figure 5.2. These include 'visual warning of approaching train', 'prompt stop/go decision' and 'dissemination of incident data'. The relationships between each of the control tasks and the situations in which they occur or could potentially occur were then mapped onto the CAT through group discussion, drawing upon the data sources described in Chapter 4.

Usually, cells surrounded by dashed lines in the CAT matrix indicate the situations where a control task *can potentially* occur given the constraints of the system, and the cells in which box and whisker symbols are displayed indicate where control tasks *typically do* occur. Empty cells indicate that the control task is not possible in that situation (Naikar et al. 2006). In our analysis, we adapted this approach by including dashed lines around cells where a control task *could* occur, given modifications to the system constraints. This enabled the analysts to consider potential redesign opportunities while conducting the analysis. For example, dashed lines were provided around the control tasks 'visual warning of approaching train' and 'audible warning of approaching train' in the situation 'road user pre-approach/rail user pre-approach', because it was considered that an in-vehicle device (system modification) could provide earlier warnings.

Several conclusions were drawn from the CATs. First, the analysis revealed very few control tasks (16%) that cannot be supported across all the situations examined (as indicated by the lack of empty cells in Figure 5.2). These cases were generally not problematic as the control tasks are not relevant to the situation (e.g. 'Exit from track' is only required when the user is on the rail level crossing itself).

Second, the analysis demonstrated how activity increases as the road user and train driver approach the rail level crossing. For example, 73% of control tasks occur when the road user is at the boom gates and the train driver is at the train detection device (the device on the track that senses the presence of the train and initiates the warning sequence), compared with only 21% of control tasks when the road user is in the pre-approach zone and the train driver is pre-whistle board.

Third, the analysis revealed that, across all situations, there are several control tasks that could occur but typically do not, as indicated by the dashed boxes in Figure 5.2; these provide opportunities for redesign. Most of these control tasks are associated with warnings of the rail level crossing or the train's approach (or related tasks including 'attract attention', 'speed reduction', 'detect train' and 'assessment of risk'), and do not typically occur in situations when the rail and road user are farthest from the rail level crossing. Examples include the following:

- Road users are not typically provided with early warnings of the rail level crossing and train. This might be achieved via an in-vehicle display or route navigation device.
- The warning system and road users are typically not aware of an approaching train until it reaches the train detection device.
- There is no information provided to road users and train drivers about the rail level crossing's risk level.

5.2.3.2 Decision Ladders

The decision ladder analysis focussed on the road user's 'stop or go' decision. This decision was defined as instances where users need to determine whether they should proceed through or stop at the crossing and wait for an approaching train to pass. The decision ladder was used to analyse the 'stop or go' decision from the point of view of different users, including drivers, pedestrians, cyclists and motorcyclists. The analysis was primarily informed by data from the diary study of rail level crossing behaviour, but for drivers was also verified using on-road data (see Chapter 4). In the diary study, 144 participants provided data regarding 457 encounters involving a train and/or activated warnings. Most encounters (92%) occurred in metropolitan Melbourne at active crossings, and the decision ladder analyses presented related to these encounters only. This included 429 encounters at 80 different crossings by 135 participants: 40 drivers (133 encounters), 33 pedestrians (128 encounters), 31 motorcyclists (86 encounters) and 31 cyclists (82 encounters).

The first component of the analysis involved developing a generic 'stop-or-go' decision ladder (see Figure 5.3), showing the range of possible decision-making processes available to different crossing users. This involved mapping data from the diary study onto the appropriate sections of the decision ladder. For example, responses to the 'situation awareness' prompt question were added to the 'Information' component of the decision ladder. The generic decision ladder therefore represents an overview of all possible decision-making processes adopted by all participants during the train encounters recorded in the diary study.

The left side of the decision ladder in Figure 5.3 summarises the various sources of information that users reported utilising to identify that they were approaching a rail level crossing (the 'alert' section of Figure 5.3), and to inform their decision to either stop or proceed through the crossing (the 'information' and 'system state' sections of Figure 5.3). Whereas many of these information sources were expected to emerge during the analysis (e.g. warning signage, active warnings, the train), there were some additional unexpected sources of information identified. These included personal triggers in the environment (e.g. a specific house or piece of vegetation) and in-vehicle alerts from a device such as a cellular phone.

When asked what the most important piece of information was in determining whether to 'stop or go', participants again described a range of information. This included the active warnings (e.g. boom gates, bells, flashing lights), behaviour of other road users, traffic lights, the location of the train, rail level crossing warning signage and seeing or hearing the train. As discussed in Chapter 4, differences across road user types were notable, as drivers and motorcyclists relied more on visual and physical warnings (e.g. flashing lights, boom gates), whereas pedestrians and cyclists relied more on auditory warnings (e.g. bells, train horn).

Moving further up the ladder, participants identified a range of options available, including proceeding through, stopping at the crossing or changing their path. The goals influencing behaviour were safety, efficiency, compliance, getting to desired destination or 'just to get through'. The right side of the decision ladder depicts the procedures available to cross and users' choice of procedure, relating back to the

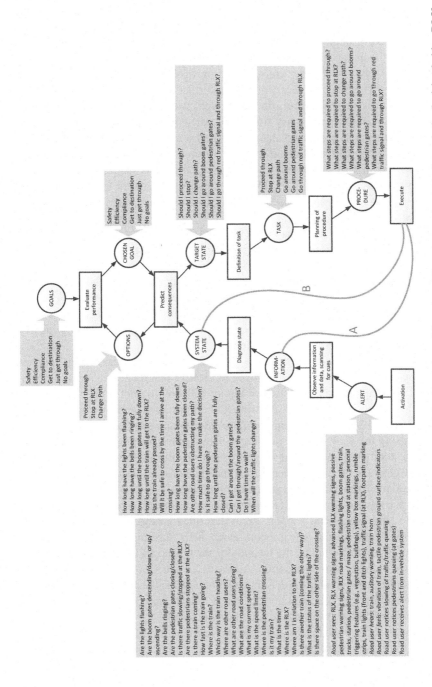

FIGURE 5.3 Decision ladder for all road users at active rail level crossings. Lines A and B represent shortcuts through the decision ladder. RLX, rail level crossing. (Adapted from Mulvihill, C.M. et al., *Appl. Ergo.*, 56, 1–10, 2016.)

options available for selection. The procedures available included to stop, proceed through, change path (to avoid the crossing), go around the boom gates, go around the pedestrian gates or cross the road at adjacent traffic lights and then cross the rail level crossing.

The generic decision ladder was used to explore differences in users' decision-making processes by overlaying the behaviour of the participants onto it. An important comparison was between compliant and non-compliant users. Thirty-one of the reported crossing traverses (6.8%) undertaken by 20 participants (13.9%) were deemed to be non-compliant (20 traverses by 11 pedestrians, 6 traverses by 5 cyclists, 3 traverses by 3 drivers and 2 traverses by 1 motorcyclist). Pedestrian participants reported a higher non-compliant behaviour rate than the other road users. The influence of goals on behaviour was apparent for the non-compliant group, with efficiency reported as the most important goal by 66.7% of non-compliant drivers and 45% of non-compliant pedestrians. For the non-compliant motorcyclist, efficiency and getting to the destination were the most important goals; safety was not identified as a goal. Safety, efficiency and getting to the destination were reported to be most important by non-compliant cyclists (33.3% each).

Analysis of shortcuts taken through the decision ladder provided some interesting insights into compliant versus non-compliant behaviour. Shortcuts were identified where the data showed that the decision-making process did not involve sequential consideration of each decision ladder element. For example, line A in Figure 5.3 represents a shortcut between 'information' and 'execute'. This means that as these users became aware of particular information in the environment, they automatically executed an action in response. This shortcut was exhibited by 73.1% of compliant drivers but no non-compliant drivers, and generally represented participants becoming aware of warnings and immediately stopping. Interestingly, line B in Figure 5.3, which represents users integrating information in the environment to come to a conclusion about the 'system state', then executing an action, was exhibited by only 13.6% of compliant drivers and 66.7% of non-compliant drivers (for more details, see Mulvihill et al. 2016). This suggests that as road users' decision-making proceeds further along the decision ladder, they are more likely to engage in non-compliant behaviour. Taken together, these shortcuts suggest that many users have well-developed skill-based and compliant responses to the onset of rail level crossing warnings. However, the system also enables users to gather information that may assist them to determine whether it is safe to engage in non-compliant behaviour. This raises interesting questions for redesign such as whether providing road users with more information may encourage non-compliance.

5.2.4 STRATEGIES ANALYSIS

The Strategies Analysis phase identifies all the different strategies that can be used to undertake the control tasks. The Strategies Analysis diagram (Cornelissen et al. 2013) was used to conduct the Strategies Analysis. This involved adding verbs (e.g. drive, ride, walk, queue, anticipate) and criteria (e.g. weather, visibility, distraction, impairment) to the abstraction hierarchy to identify the range of strategies available to different rail level crossing users. The strategies were identified from data collected in the on-road studies and diary study, and also from formatively identifying pathways within the abstraction hierarchy itself. Examples of the verbs and criteria used are overlaid onto the abstraction hierarchy for active rail level crossings in Figure 5.4. In addition, an example strategy identified in the Strategies Analysis is highlighted with bolded means-ends links. The strategy highlighted is for a car driver and, reading from the bottom of Figure 5.4 to the top, indicates that drivers 'brake' (verb) their 'vehicles' (physical object), which affects 'locomotion' (physical functionality) and contributes to the functions of 'behave appropriately for the environment' and 'maintain road and rail user separation'. This strategy is likely to be undertaken when 'warnings present', 'warnings active' and in situations of 'low visibility' (criteria). Finally, the strategy contributes to the functional purposes of 'protect road users' and 'protect rail users'.

Multiple possible strategies for negotiating rail level crossings were identified for each type of road user. For example, car drivers can become aware of an approaching train via multiple physical objects, including the train itself, warning devices (e.g. flashing lights, bells, boom barriers) or the behaviour of other road users (e.g. lead vehicles stopping). This highlights high levels of flexibility and redundancy within current rail level crossing systems. In addition, when this is combined with the decision ladder analysis, it becomes clear that, even within specific road user groups, there is no single consistent way of negotiating rail level crossings.

An interesting finding from the Strategies Analysis was that warnings signalling the presence of the crossing (e.g. advanced signage) do not appear to be highly relevant or well used. Rather, active warnings of an approaching train and the rail level crossing infrastructure itself were more prevalent in the strategies identified. This finding questions the use of some specific passive warnings within active rail level crossing environments – particularly advanced warning signage situated well ahead of the rail level crossings. This notion is enhanced by urban on-road study findings that drivers are not always aware that they have passed through a rail level crossing (e.g. Salmon et al. 2013a, Young et al. 2015). In combination with the findings from the CAT analysis, it suggests that although warnings of an upcoming rail level crossing *could* be provided earlier on the road user's approach, this needs to be carefully designed to ensure it will integrate with the strategies road users employ.

The nature of information provided to users about approaching trains is also brought into question by the Strategies Analysis. It seems that the absence of specific information regarding the approaching train is an important constraint. Currently, the information provided is in the form of generic barriers and warnings (i.e. 'a train is coming'). The Strategies Analysis suggests that many strategies identified would be better supported through the provision of more specific information, such as the number and direction of trains approaching, and potentially information regarding time-to-arrival and expected delay. Notably, the analysis identified a range of strategies currently employed by users attempting to seek this information. For example,

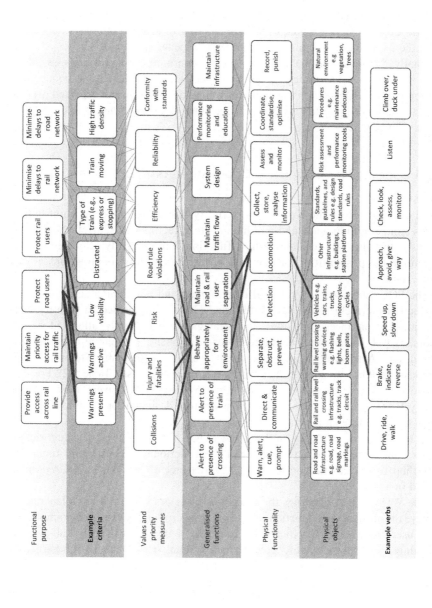

FIGURE 5.4 Strategies analysis diagram for active rail level crossings. (Adapted from Salmon, P.M. et al., *Ergonomics*, 12, 1–10, 2016.)

pedestrians look down the track and attempt to predict the time remaining before the train arrives. This information is then used to determine whether they should cross, regardless of the state of the warnings.

Potential conflicts between road user types employing different strategies were also identified. For example, pedestrians may traverse the crossing via the road instead of the footpath, thus coming into conflict with drivers attempting to traverse the rail level crossing. Further, cyclists attempting to avoid boom gates may cross using the footpath, thus coming into conflict with pedestrians.

Finally, the physical use of infrastructure by non-motorised road users was identified in many crossing strategies. For example, cyclists use fencing or boom gates to support their balance while waiting for trains to pass. These unexpected uses of infrastructure are important as they provide clues regarding features that could be included in new designs.

5.2.5 SOCIAL ORGANISATION AND COOPERATION ANALYSIS

The SOCA phase of CWA examines which actors currently do what and who could do what, given the constraints of the system. The important point is that actors can be both humans and non-humans (e.g. technologies, artefacts). In this analysis, the aim was to identify how functions are currently allocated and to determine the possibilities for new and different allocation of functions at rail level crossings.

The SOCA involved mapping different human and non-human actors onto the abstraction hierarchies, decision ladder and CATs. This helped to identify how functions, affordances, decisions and strategies are currently allocated across actors, as well as how they could be allocated to identify potential opportunities for redesign. Actors were classified into two levels – category sub-systems and their composite actors – to support differing levels of analysis (see Table 5.2). For this categorisation, road users were separated into vehicle users (drivers, motorcyclists and cyclists) and pedestrians, as the infrastructure designed for these users is different (road vs. footpath).

In this chapter, we focus on the WDA SOCA that shows which categories of actors contribute to the achievement of the functional purposes and functions identified in the abstraction hierarchy. An extract of the WDA SOCA is presented in Table 5.3.

The WDA SOCA identifies the categories of actors who need to collaborate to achieve the purposes and functions of the rail level crossing system. For example, the 'road user sub-system' (specifically the road itself), 'detection and alert sub-system' (warnings) and 'infrastructure provider sub-system' (road and rail infrastructure owners) currently contribute to the functional purpose 'provide access across rail line'. For the function, 'alert to presence of rail level crossing', the following categories of actors currently contribute: the 'detection and alert sub-system', 'regulators/authorities sub-system' (through providing road and rail infrastructure) and 'physical infrastructure sub-system' (the rail level crossing itself).

TABLE 5.2

Actors Considered in the SOCA Phase

Category	Actors
Rail user sub-system	Train driver
	Train
	Train tracks
Road user sub-system	Driver/motorcyclist/cyclist
	Road vehicle
	Road
Pedestrian sub-system	Pedestrian
	Footpath
Detection and alert sub-system	Active warning systems
	Signage
	Detection systems
Physical infrastructure sub-system	Physical infrastructure (infrastructure at the rail level crossing itself)
Regulators/authorities sub-system	Road authority
	Rail regulator
	Government/policy makers
	Police
Infrastructure provider sub-system	Rail infrastructure owner/maintainer
	Road infrastructure owner/maintainer
Train service provider sub-system	Train service provider
Media sub-system	Media

The formative element of the SOCA is also included in Table 5.3, via the actors presented in bold text. This shows how existing groups of actors could contribute to the functional purposes and functions if modifications to system constraints were made. For example, the 'road user sub-system' (specifically the road vehicle) has been bolded for the function, 'alert to presence of rail level crossing' as an in-vehicle display, mobile phone or GPS route navigation system could potentially provide a warning of an upcoming rail level crossing. Likewise, for the 'system performance monitoring and education' function, the 'rail user sub-system', 'road user sub-system', 'pedestrian sub-system' and 'detection and alert sub-system' have been included as, with the provision of appropriate reporting mechanisms, actors in these groups could conceivably collect and report information regarding how rail level crossing systems are functioning (e.g. incidents and near misses, failed components).

The SOCA demonstrated various opportunities for reallocating functions within rail level crossing systems and for adding redundancy by increasing the number of actors performing functions within the system. A notable finding is the existing design's heavy reliance on non-human and rail level crossing-related actors to achieve functions (e.g. signage, warnings, barriers, trains), leaving road users such as drivers and vehicles under-utilised outside of the 'stop or go' decision.

TABLE 5.3
Extract from the Social Organisation and Cooperation Analysis

Nodes from Abstraction Hierarchy	Contributing Actor Categories
Functional purposes	
Provide access across rail line	• Road user sub-system • Detection and alert sub-system • Physical infrastructure sub-system
Maintain priority access for rail traffic	• Detection and alert sub-system • Regulators/authorities sub-system • Physical infrastructure sub-system
Protect road users	• Rail user sub-system • Road user sub-system • Pedestrian sub-system • Regulators/authorities sub-system • Detection and alert sub-system • **Train service provider sub-system** • Media sub-system • Physical infrastructure sub-system
Protect rail users	• Rail user sub-system • Road user sub-system • Pedestrian sub-system • Regulators/authorities sub-system • Detection and alert sub-system • Train service provider sub-system • **Physical infrastructure sub-system**
Minimise delays to road network	• **Rail user sub-system** • Regulators/authorities sub-system • Detection and alert sub-system • **Train service provider sub-system**
Minimise delays to rail network	• **Road user sub-system** • Regulators/authorities sub-system • Detection and alert sub-system • Train service provider sub-system
Functions	
Alert to presence of rail level crossing	• **Road user sub-system** • Regulators/authorities sub-system • Detection and alert sub-system • Physical infrastructure sub-system
Alert to presence of train	• Rail user sub-system • **Road user sub-system** • Regulators/authorities sub-system • Detection and alert sub-system

(Continued)

TABLE 5.3 (*Continued*)
Extract from the Social Organisation and Cooperation Analysis

Nodes from Abstraction Hierarchy	Contributing Actor Categories
Behave appropriately for environment	• Rail user sub-system • Road user sub-system • Pedestrian sub-system • Regulators/authorities sub-system • Detection and alert sub-system • Media sub-system
Maintain road user and rail separation	• Rail user sub-system • Road user sub-system • Pedestrian sub-system • Detection and alert sub-system • Physical infrastructure sub-system
Maintain road user/rail/pedestrian flow	• Rail user sub-system • Road user sub-system • Pedestrian sub-system • Detection and alert sub-system • Physical infrastructure sub-system
System design	• Regulators/authorities sub-system • Train service provider sub-system
System performance monitoring and education	• **Rail user sub-system** • **Road user sub-system** • **Pedestrian sub-system** • **Detection and alert sub-system** • Regulators/authorities sub-system • Train service provider sub-system • Infrastructure provider sub-system • Media sub-system
Maintain infrastructure	• Regulators/authorities sub-system • Infrastructure provider sub-system • **Physical infrastructure sub-system**

Note: Actors who could potentially perform functions and functional purposes but currently do not are indicated in bold.

5.2.6 Summary of Findings from CWA

Several key findings were identified from each CWA phase. For the WDA phase, the abstraction hierarchy showed the following:

- Conflicting purposes of the system.
- Values and priorities that are not currently being achieved.

- The requirement for new data systems to support better understanding of how the system is meeting its purposes.
- A range of ways in which functions can fail and lead to collisions.

For the ConTA phase, the CAT and decision ladders identified the following:

- Multiple factors influencing road user decision-making.
- The diverse situation awareness requirements for different users.
- The range of goals impacting road user behaviour.
- The options available to road users when negotiating crossings.
- Information processing shortcuts that road users take to make the decision to 'stop or go'.
- Unexploited opportunities for control tasks to be performed in new situations.

In the Strategies Analysis phase, the use of the strategies analysis diagram led to the following key findings:

- There are a range of ways in which different users can negotiate rail level crossings.
- Strategies employed by different users can conflict.

Finally, the SOCA phase identified the following:

- An over-reliance on particular actors and artefacts, particularly the rail level crossing warnings.
- Potential new allocations of functions across actors in the rail level crossing system, for example, road vehicles represent one opportunity to provide additional warnings of an approaching train.

5.3   HTA OF RAIL LEVEL CROSSING SYSTEMS

As outlined in Chapter 2, HTA (Annett 2004) is primarily concerned with goals (an objective or end state) and their decomposition into sub-goals and the requisite physical and cognitive operations (Annett and Stanton 1998). HTA works by decomposing systems into a hierarchy of goals, sub-ordinate goals, operations and plans. It focusses on 'what an operator ... is required to do, in terms of actions and/ or cognitive processes to achieve a system goal' (Kirwan and Ainsworth 1992, p. 1). Note that an 'operator', like an actor in CWA, may be a human or a technological operator (e.g. equipment, devices and interfaces). HTA outputs therefore specify the overall goal of a system, the sub-goals required to achieve this goal, the operations necessary to achieve each of the sub-goals specified and the plans that describe the ordering of sub-goals and operations.

HTA has previously been undertaken to describe the full driving task (see Walker et al. 2015). The analysis undertaken for the current project therefore focussed specifically on drivers' interactions with rail level crossings.

5.3.1 Analysis Approach

The aim of the HTA was to produce a goal-based description of drivers' interactions with passive and active rail level crossings. Although this is useful alone, the intention was to produce an HTA description of rail level crossing functioning that could then be used to inform a human error identification process (specifically, SHERPA; Embrey 1986).

Multiple analysts with significant experience in applying HTA to diverse domains (e.g. defence, road and rail transport, aviation, maritime) were involved in constructing the HTA. The data inputs were the same as those used to generate the WDA (see Table 5.1).

5.3.2 HTA of Rail Level Crossings

Extracts from the HTA are presented in Figures 5.5 and 5.6. The plans shown are defined for interaction with an active crossing in a rural area; however, the plans can be adapted so that the HTA describes an urban active crossing, or a passive crossing, by leaving out operations that are not applicable. For example, for a passive crossing, the only operation supporting sub-goal 5 'Announce presence of train' would be 5.5 'Sound horn (Train driver)' as there are no other active warnings at passive rail level crossings.

Figure 5.5 shows the nine sub-ordinate goals underpinning the main goal of 'Safe and efficient de-confliction of road and rail traffic'. Assuming the driver maintains control of the vehicle throughout, the sub-ordinate goals initially include the following three key detection tasks:

1. The rail level crossing warning system detecting the presence of the approaching train
2. The vehicle driver and train driver detecting the presence of a rail level crossing
3. The train driver detecting the presence of road users at the crossing

Following detection of the train, the rail level crossing warning system then announces the presence of a train through active warnings such as flashing lights. The onus is then placed on the driver to detect the presence of the train (either through the warnings or through seeing the train) and stop appropriately. Once the train traverses the rail level crossing, the warning system deactivates and the driver can proceed.

The sub-ordinate goals are decomposed in Figure 5.6 to reveal sub-goals and associated operations. The description of the system goals and sub-goals provided by the HTA demonstrates the amount of activity and interdependence of activity

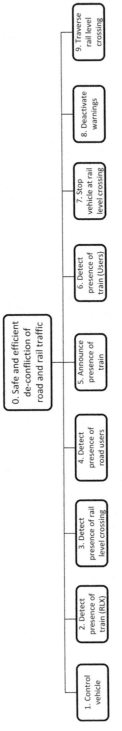

FIGURE 5.5 Rail level crossing HTA super- and sub-ordinate goals. RLX, rail level crossing.

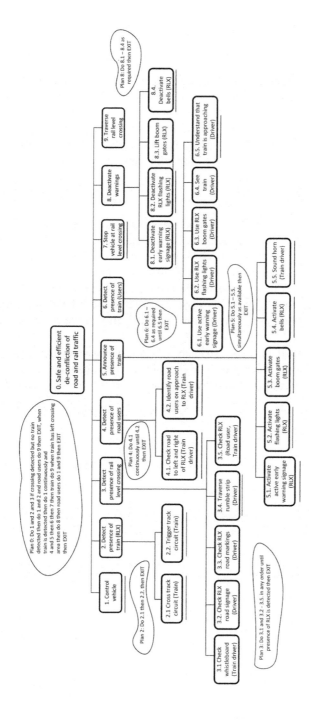

FIGURE 5.6 Rail level crossing (RLX) sub-goal decomposition. Note: Nodes without a line underneath them are decomposed further in the full analysis.

occurring at a rail level crossing. This demonstrates the flexibility of HTA beyond considering the road user perspective alone.

5.4 SHERPA OF RAIL LEVEL CROSSING SYSTEMS

As described in Chapter 2, SHERPA (Embrey 1986) is a human error identification method used to identify the potential errors that could arise during task performance. In the current project, SHERPA was applied to identify the errors that drivers could make at existing rail level crossings in Victoria, Australia. In addition, we were interested in understanding the error tolerance of current rail level crossing designs and to identify initial design ideas for improving behaviour and safety.

5.4.1 ANALYSIS APPROACH

Initially, one analyst used SHERPA to identify potential errors. This involved taking each bottom-level task step from the HTA, classifying it into one of the five SHERPA behaviours (Action, Check, Retrieval, Communication, Selection) and then using the error mode taxonomy in Figure 5.7 to identify credible errors. For each credible error, a description of the error and its consequences were documented along with any recovery steps (i.e. the point in the HTA at which the error could be recovered), ratings of probability and criticality of the error and potential remedial measures.

Following the initial analysis, two analysts with experience in applying SHERPA reviewed the analysis to check the credibility of the errors and their associated probability and criticality ratings. An extract of the rail level crossing SHERPA is presented in Table 5.4.

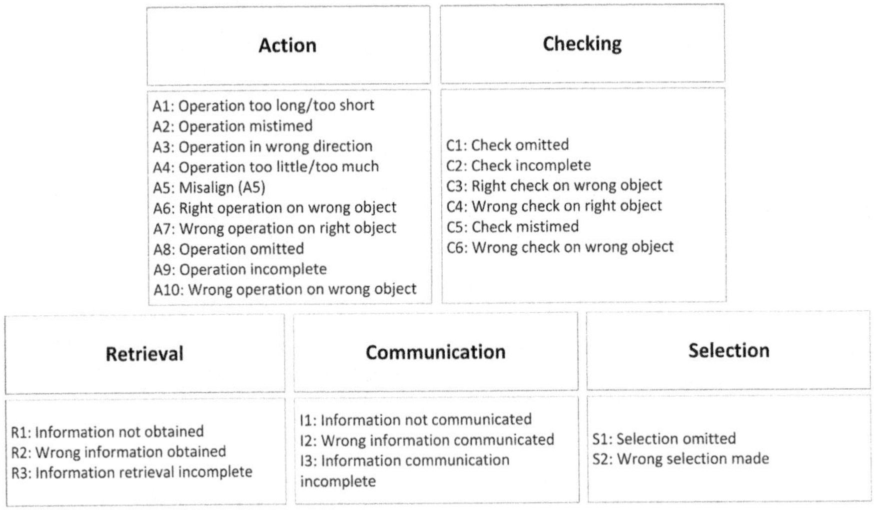

Action	Checking
A1: Operation too long/too short A2: Operation mistimed A3: Operation in wrong direction A4: Operation too little/too much A5: Misalign (A5) A6: Right operation on wrong object A7: Wrong operation on right object A8: Operation omitted A9: Operation incomplete A10: Wrong operation on wrong object	C1: Check omitted C2: Check incomplete C3: Right check on wrong object C4: Wrong check on right object C5: Check mistimed C6: Wrong check on wrong object

Retrieval	Communication	Selection
R1: Information not obtained R2: Wrong information obtained R3: Information retrieval incomplete	I1: Information not communicated I2: Wrong information communicated I3: Information communication incomplete	S1: Selection omitted S2: Wrong selection made

FIGURE 5.7   SHERPA error mode taxonomy (Embrey, 1986).

TABLE 5.4

Extract of SHERPA for Driver Behaviour at Rail Level Crossings

Task	Error Mode	Error Description	Consequence	Recovery	P	C	Potential Remedial Measures
6.1.1. Look for early warning signage	C5	Driver looks for signage too late	Driver fails to comprehend approaching train	Step 6.2.1	H		In-vehicle reminder system Runway red lights in stop line (in-road studs) Alternating LED image of moving train on active early warning signage (or replace current signs with this) Traffic lights linked to RLX
6.1.2. Detect flashing lights	R1	Driver fails to detect flashing lights	Driver fails to comprehend approaching train	Step 6.3.1	M	✓	In-vehicle system Runway red lights in stop line (in-road studs) Alternating LED image of moving train on active early warning signage (or replace current signs with this) Traffic lights linked to RLX
6.1.2. Detect flashing lights	C5	Driver detects flashing lights too late	Driver fails to comprehend approaching train	Step 6.3.1	M	✓	In-vehicle system Runway red lights in stop line (in-road studs) Alternating LED image of moving train on active early warning signage (or replace current signs with this) Traffic lights linked to RLX
6.1.3. Interpret flashing lights	R1	Driver fails to interpret flashing lights (LBFTS)	Driver fails to comprehend approaching train	Step 6.3.1	M	✓	In-vehicle system Runway red lights in stop line (in-road studs) Alternating LED image of moving train on active early warning signage (or replace current signs with this) Traffic lights linked to RLX

C, criticality; H, high; L, low; M, medium; P, probability; RLX, rail level crossing; LBFTS, looked-but-failed-to-see error.

5.4.2   SHERPA ANALYSIS OF RAIL LEVEL CROSSINGS

A total of 92 potential errors were identified, which were fairly evenly distributed across three categories: action errors (e.g. driver fails to slow, rail level crossing fails to activate warnings), checking errors (e.g. driver fails to look at flashing light assembly, train driver fails to look for road users) and retrieval errors (e.g. driver fails to interpret flashing lights, driver misreads signage).

The following tasks had the greatest number of potential errors associated with them:

- Detect the presence of train (road vehicle driver)
- Detect rail level crossing (road vehicle driver)

Moreover, a greater proportion of potential errors were associated with road vehicle drivers, as opposed to the train driver or the technical components of the rail level crossing (e.g. train detection device, flashing lights, boom gates). This reflects the high onus placed on the human user, consistent with the findings from the CWA. An important design implication is that new designs should exploit other components within the system such as the vehicle, technology-based detection systems and the train itself.

A final notable feature of the SHERPA analysis was that there appears to be few redundancies or opportunities for error identification, recovery and mitigation in current rail level crossing environments to cope with high criticality errors such as the road user failing to detect flashing lights. This is especially the case with rail level crossings that do not currently have boom gates.

5.5   SUMMARY

The aim of this chapter was to present the findings derived from a systems analysis of rail level crossings that involved applying CWA, HTA and SHERPA to active and passive rail level crossings in Victoria, Australia. The tripartite analysis provided an important insight into the functioning of rail level crossing systems, as well as the issues currently threatening safety, and factors to be considered in new rail level crossing designs.

Across these analyses, several key risks were identified, mostly from the decision ladder analysis, Strategies Analysis and the SHERPA application. These risks were as follows:

- Road users not aware of an upcoming rail level crossing
- Road users not aware of rail level crossing warnings
- Road users not aware of an approaching train
- Road user not checking for trains sufficiently
- Road user not detecting a second or subsequent train

- Road user misjudging the speed or distance of the train
- Road user choosing to cross when warnings are activated/a train is approaching
- Road user queuing or short stacking on the crossing
- Warning systems failing to announce the presence of a train

In addition, several factors were identified that contribute to and escalate the likelihood of these risks, the key factors were as follows:

- Visual clutter and distractions in environment (especially in urban environments)
- Monotonous environments (e.g. in rural environments) and task under-engagement
- Congestion
- Psychological reluctance to stop
- Frustration
- Expectancy
- Time pressure
- Panic

Taken together, the analyses clearly demonstrate the inherent complexity within rail level crossing systems. They have multiple competing purposes, many values and priorities, various constraints on behaviour and multiple pathways to failure. In addition, different users have various strategies available to them, meaning there is no 'one-size-fits-all' approach in terms of the information they use and activities they perform. These differences occur both across user groups (e.g. drivers vs. pedestrians) and within user groups (e.g. differences between drivers), and relate to the sources of information used, the goals pursued and the courses of action employed. There are multiple technological and human agents performing key tasks, some of which occur away from the rail level crossing in time and space yet can be critical to performance. Because of this complexity, there are many design-induced errors that can potentially threaten safety.

The implication of the analysis is that change may not only be required at the rail level crossings themselves (e.g. introducing new ways of alerting road users to the presence of a train), but that fundamental changes may also be required throughout the overall road and rail system. For example, the presence of competing functional purposes represents a barrier to implementing designs focussed purely on improving safety as they may adversely impact other functional purposes (e.g. efficiency). Designs in which trains slow or stop at rail level crossings are used in Europe and appear to have a safety benefit; however, with a need to maintain efficiency over long geographical distances, it is questionable whether such approaches would be practical in Australia. This demonstrates the need to consider the rail level crossing system not just as the intersection of the road and rail transport systems, but as part of an overall transport system that itself contributes to wider social, economic and environmental goals. Other systemic issues highlighted by the analysis include the following:

- A requirement for improved data collection and analysis systems (e.g. near miss and violation reporting systems)
- A requirement for improved risk assessment systems that consider the inherent complexity underpinning behaviour at rail level crossings
- A requirement for a modified design process that enables flexibility outside of the current 'design to meet standards' approach.

The findings also led to a number of insights that can inform the design of future rail level crossing systems. These are discussed further in Chapter 6.

Section III

Design of New Rail Level
Crossing Environments

6 A Participatory Approach to Designing Rail Level Crossings

With Contributions from:
Christine Mulvihill and Kristie Young

6.1 INTRODUCTION

This chapter describes the process we used to generate new rail level crossing design concepts. Although the utility of Cognitive Work Analysis (CWA) for understanding complex system performance is well established, its outputs are not often used directly in the design process, particularly where the scope is broader than interface design (Read et al. 2012). It has been proposed that CWA provides 'a philosophical tool' rather than a design method (Mendoza et al. 2011, p. 58). Lintern (2005) explains that CWA provides recommendations for system design; it is up to designers to decide how these will be implemented. To explore this further, we conducted a survey of CWA users to better understand current practice and found that there appears to be no standard approach to whole of system design with CWA; instead, practitioners often craft their own approaches (Read et al. 2015a). In response to these concerns and to support the development of new rail level crossing design concepts, we developed the CWA Design Toolkit (CWA-DT; Read et al. 2015c). The intent was not to replace the CWA framework, but to extend it with tools that can be used to translate CWA outputs into designs.

6.2 PHILOSOPHY UNDERPINNING THE CWA-DT

The CWA-DT was developed to support the generation of new designs underpinned by the values and principles of sociotechnical systems theory. It is important that these values (see Box 6.1) underpin both the *process* of doing design and the *outcome* of design. Accordingly, during the design process, human involvement and creativity should be embraced, as it should in the designs produced.

> **BOX 6.1   VALUES OF SOCIOTECHINCAL SYSTEMS THEORY (SEE CHAPTER 3)**
>
> - Humans as assets
> - Technology as a tool to assist humans
> - Promotion of quality of life
> - Respect for individual differences
> - Responsibility to all stakeholders

Further, the content principles of sociotechnical systems theory (see Figure 6.1) should be evident in the outcomes of the design process. The tools and approaches within the CWA-DT were developed to encourage the incorporation of sociotechnical systems theory values and process principles (see Chapter 3) to create designs that meet the content principles.

The principles can be elaborated as follows:

- *Tasks are allocated appropriately between and among humans and technology*: Design of work and equipment should be based on the complementarity of people and machines, not on competition between them. Recognition of the unique capabilities of humans as adaptive elements in sociotechnical systems is essential to the design and effective functioning

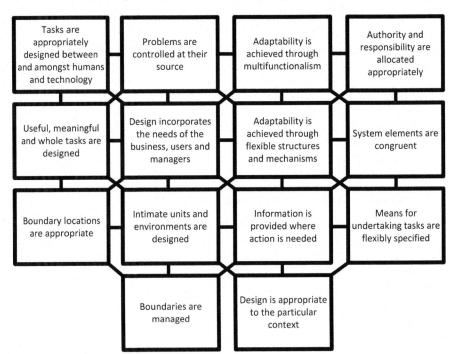

FIGURE 6.1   Sociotechnical systems theory content principles.

of organisations (Davis 1982). Tasks involving unpredictable contexts and judgement should be allocated to humans rather than computers (Clegg 2000).

- *Useful, meaningful, and whole tasks are designed*: A process perspective should be taken, and people should be given the authority and resources to conduct, supervise or manage complete processes. Work should incorporate whole tasks, rather than fragmented parts (Clegg 2000).
- *Boundary locations are appropriate*: Although boundaries exist between functional units or groups, core processes should not be split across these artificial boundaries (Clegg 2000). Further, boundaries should not be drawn that impede communication and the sharing of knowledge, experience and learning (Cherns 1987).
- *Boundaries are managed*: The focus of a supervisor should be the management of boundary activities, such as ensuring the group has sufficient resources, is coordinating with other groups and predicting changes that may impact on their activities (Cherns 1976). Supervisors provide the work group with a buffer to protect them from external disturbances or changing demands, and support them through provision of resources and training (Davis 1982). Boundaries may be created and managed by the work group itself, by defining the rules and processes that constrain their activities and by participating in the ongoing revision and update of rules (Hirschhorn et al. 2001).
- *Problems are controlled at their source*: Problems or variances (i.e. unplanned events that critically affect objectives) should be controlled as near to their point of origin as possible (Cherns 1976). Designing to promote learning from mistakes or variances is beneficial (Sinclair 2007). Users should have access to the means for controlling or resolving problems themselves and the necessary authority or competency to do so (Davis 1982).
- *Design incorporates the needs of the business, users and managers*: Consideration should be given to the objectives of the overall system (e.g. the goals of the business) as well as to the goals of employees, managers and any other end users (Clegg 2000).
- *Intimate units and environments are designed*: Organisational and physical structures should provide small, intimate environments for individuals and groups (Davis 1982).
- *Design is appropriate to the context*: Design choices do not necessarily have universal applicability. Choices should be based on the context of operations, the goals to be achieved and the local context of implementation (Clegg 2000). This requires an individualised design to suit the specific situation rather than simply copying solutions that have been implemented elsewhere.
- *Adaptability is achieved through multifunctionalism*: Systems must adapt to ever changing environments. The system needs a repertoire of responses to environmental changes (Cherns 1976, 1987). Ashby's Law of Requisite Variety proposes that it is people who have the knowledge, flexibility and

agility to provide adaptive responses, which facilitate organisational resilience (Sinclair 2007). Resilience can be achieved by training and multi-skilling people to enable them to perform multiple functions and to assist in tasks outside of their normal area during unpredictable events or emergencies, or when change occurs (Cherns 1976). It may not be possible or desirable for all individuals to have all skills. In such cases, dynamic collaboration is needed between those with complementary knowledge and experiences (Hirschhorn et al. 2001).

- *Adaptability is achieved through flexible structures and mechanisms*: Self-maintaining units or groups should be implemented which have the capability to perform all the activities required to achieve its specific goals under a wide variety of contingencies. Units should be flexible and able to adapt to changing environmental demands (Davis 1982).
- *Information is provided where action is needed*: Information systems should be designed to provide information at the time and place where action will be required (Cherns 1976). They should provide information to those taking action rather than those controlling or monitoring the system (Davis 1982) and should give the right type and amount of feedback to users to enable them to learn from experience and to anticipate events (Cherns 1976).
- *Means for undertaking tasks are flexibly specified*: Although it is important for design to specify what is essential, no more than what is necessary should be specified. Design should make it clear what needs to be achieved (the ends), but it is often unnecessary to specify how it must be achieved (the means; Cherns 1976). Design should avoid over-specifying how tasks must be performed, as this limits adaptability. Users, as local experts, should be allowed to solve their own problems and develop their own methods of working, thereby incorporating scope for learning and innovation (Clegg 2000).
- *Authority and responsibility are allocated appropriately*: Those who require access to equipment and resources should have the authority to access and command them. In return, they must be accountable for their use (Cherns 1987). Further, in allocating responsibility, differences in status and privilege should be minimised where they are unnecessary to the achievement of goals (Davis 1982). Finally, authority should be distributed so that people doing the work are empowered to make appropriate decisions according to their role (Sinclair 2007).
- *System elements are congruent*: A new design needs to be congruent with surrounding systems and practices to facilitate implementation. However, it may be that a new system requires some accommodation by the systems into which it is being placed. It may prove to be a catalyst for change (Clegg 2000). Further, congruence needs to be achieved between the work design and the overall goals of the system (Sinclair 2007).

6.2.1 Contrasting Sociotechnical Systems Theory and Traditional Safety Management Approaches

Because of these principles, sociotechnical systems theory-based designs are typically quite different to designs derived from traditional approaches. In the case of transportation, designs tend to follow a safety management approach based on the notion of risk control, using frameworks such as the hierarchy of control. Figure 6.2 provides an indicative distribution of intervention types in terms of their alignment with sociotechnical systems theory or with traditional safety management approaches. Sociotechnical systems theory approaches attempt to enhance flexibility and local adaptation through giving users more control and latitude for behaviour, focussing on supporting users to achieve their goals and ultimately placing design in the hands of users. In contrast, the safety management approach seeks to separate users from the hazard (the train) by using physical constraints (e.g. barriers) and focusses on achieving compliance through enforcement, education and rewards. In between these extremes, there are interventions that present a hybrid approach, such as providing better information to users or giving users a sense of meaning from their interaction with the design.

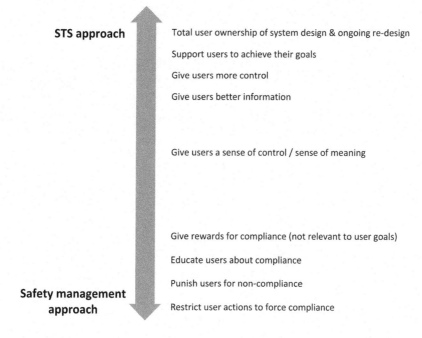

FIGURE 6.2 Continuum of interventions aligned with sociotechnical systems theory versus traditional safety management approaches.

6.3   APPLICATION OF THE CWA-DT TO RAIL LEVEL CROSSING DESIGN

Phase 3 of the research program involved the development of novel designs for urban and rural rail level crossings. It was driven by the application of the CWA-DT as part of a participatory design process involving rail and road safety stakeholders (see Figure 6.3 for an overview of the process).

The remainder of this chapter describes the application of the CWA-DT to generate initial design concepts for rail level crossings.

6.3.1   DOCUMENTATION OF INSIGHTS FROM THE CWA OUTPUTS

The CWA-DT places importance on the translation of 'insights', gained from CWA analyses, into design solutions. Insights include both non-obvious inferences from the analysis outputs and more obvious findings about the system that the research

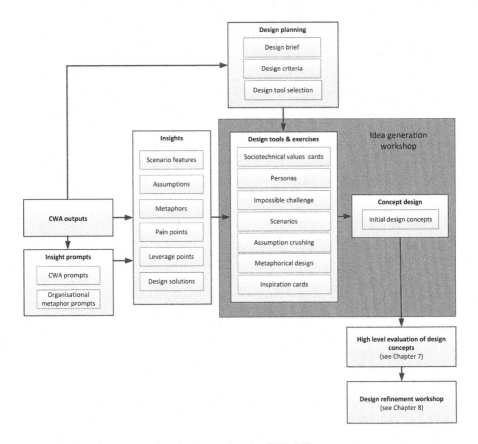

FIGURE 6.3   Key steps undertaken following the CWA-DT process.

team considers important for the design process. The categories of insights used in the CWA-DT are described in Box 6.2.

Insights were documented during the data collection and analysis processes, and during the design planning process. Examples of the insights identified are provided in Table 6.1.

BOX 6.2   CWA-DT INSIGHT CATEGORIES

- *Assumptions*: Assumptions relate to the underlying hypotheses, expectations and beliefs upon which the system, or part of the system, is based. They could be about the way the system functions or how people are expected to behave within the system. They could be correct or incorrect.
- *Leverage points*: Aspects within a system that, if changed in a small way, could produce larger changes across the system. For example, there may be evidence within the analysis that suggests there is an actor or process that is under-utilised or could be better utilised to meet the systems' purpose(s).
- *Metaphors*: Commonly used in design, metaphors and analogies promote thinking about how to apply existing ideas in new situations. Metaphor involves the comparison, interaction or substitution of two subjects on a symbolic level. An insight might involve, for example, realising that there are similarities in two domains (i.e. scheduling in manufacturing and health care) or that something in the natural environment is similar to what is trying to be achieved through technology (i.e. comparing an aircraft wing and a bird's wing).
- *Pain points*: Problems or issues identified during the analysis. They may be points of frustration for users, conflicting goals between users or problems such as information bottlenecks in organisational systems.
- *Scenario features*: Data collection and analysis activities often uncover rich contextual information about the domain being analysed. The intention of this insight type is to capture the key features that would be important to consider in the design process. A feature of a potential scenario could include a type of actor, attributes of an actor, a type of task or an environmental disturbance or influence.
- *Design solutions*: A proposed design or feature of a design identified by an analyst or others reviewing the analysis outputs. The solution does not have to be a well-developed idea to be documented.

TABLE 6.1

Examples of Insights that Informed the Design Process

Insight	Insight Type	CWA Phase
Train drivers currently have very few options for avoiding a collision: they can sound the train horn and activate the emergency brakes, but in almost all cases the train will not be able to stop in time to avoid a collision. This lack of control over the situation can add to the distress and trauma experienced by train drivers. As sociotechnical systems theory notes, control is very important. How can we give train drivers greater control to manage these situations?	Leverage point	Decision ladder
Expectancy plays a key role in road user behaviour at rail level crossings. Drivers who experience rail level crossings with no trains present are likely to not expect to see a train, and therefore may not look for trains.	Pain point	Decision ladder and addition data analysis from rural on-road study
Rain and wet weather influences behaviour, particularly of pedestrians (e.g. running in front of a train rather than waiting in the rain).	Scenario feature	Strategies analysis diagram

6.3.2 PROMPTING FOR INSIGHTS

To assist in the identification of insights, the CWA-DT includes a series of prompt questions as well as a template to assist researchers in documenting insights. These can be used to generate further insights that did not occur spontaneously during the analysis process. Two sets of prompt questions are provided to assist researchers to interrogate the CWA outputs for insights, particularly in team design situations where not all individuals were involved in developing all outputs. It is recommended that a selection of the prompts be applied, depending upon the aims and scope of the research project being undertaken. The prompts are intended to be used in facilitated group settings to encourage systems thinking in design.

The first set of prompts used in the CWA-DT relates to the different phases of CWA. These 'CWA prompts' aim to draw out relevant findings and insights based on the CWA literature (e.g. descriptions offered by Jenkins et al. [2009] and Vicente [1999]) and pose questions drawn from the sociotechnical systems theory literature, particularly those offered by Appelbaum (1997).

Examples of the prompt questions include the following (for a full list, see Read et al. [2016b]):

- What is the system's greatest strength? (all phases)
- What is the system's most obvious weakness? (all phases)

- Are there multiple purposes specified for the system? Do they conflict? Could they potentially conflict? Under what circumstances? (Work Domain Analysis)
- Are any purposes of the system not well supported? (Work Domain Analysis)
- For what situations it is possible to complete tasks, although they are not typically undertaken? Why are they not typically undertaken? (Control Task Analysis)
- Are there situations involving high workload (many functions typically performed)? (Control Task Analysis)
- Which strategies are reinforced or rewarded within the system? (Strategies Analysis)
- Which strategies are not rewarded or are punished within the system? (Strategies Analysis)
- To what extent are tasks currently completed by humans? To what extent are tasks currently completed by technology? (Social Organisation and Cooperation Analysis)
- Would any tasks completed by humans be better completed by technology or vice versa? (Social Organisation and Cooperation Analysis)

The second set of prompts takes a somewhat different approach. The 'Organisational metaphor prompts' aim to promote innovative or 'out of the box' thinking about the system by challenging assumptions about how organisations and other complex systems operate. The prompts are based upon four paradigms or world views proposed by Morgan (1980), through which organisational functioning can be viewed. The paradigms are described in Box 6.3. The paradigms vary regarding the extent to which they align with ideas of control and regulation versus openness to radical change. They also vary regarding the extent to which they are concerned with objective or subjective views of system functioning. Within each paradigm, Morgan's (1980) metaphors for organisational functioning can be used to expand thinking about organisations and systems. For example, in the functionalist paradigm, the cybernetic metaphor is interested in patterns of information and how feedback loops enable learning. In the radical humanist paradigm, the metaphor of the psychic prison, which draws upon Marxist theory, involves an analysis to identify the properties of a system that are alienating or that dominate actors' activities.

We used these metaphors as a basis for developing the second set of prompt questions. Some example questions include the following (see Read et al. [2016b] for the full list):

- *Loosely coupled systems metaphor*: Where is coordination between system components, actors or groups of actors unsuccessful or lacking?
- *Population ecology metaphor*: What foreseeable environmental changes could affect the system's position relative to its competitors?
- *Accomplishments metaphor*: What are the social rules or patterns that assist actors to successfully interact within the system?

BOX 6.3 PARADIGMS AND METAPHORS UNDERPINNING THE ORGANISATIONAL METAPHOR PROMPT QUESTIONS

- *Functionalist paradigm*: The functionalist paradigm assumes there is an objective reality, which is functionally structured and can be controlled to ensure goals are met and maintained. This aligns well with the theoretical underpinnings of CWA and sociotechnical systems theory. Metaphors within this paradigm include Culture, Theatre, Political systems, Loosely coupled systems, Cybernetic systems, Population ecology, Organism and Machine.
- *Interpretive paradigm*: This paradigm emphasises that reality is socially constructed and is the product of the subjective and inter-subjective experiences of individuals. It focusses on the processes by which realities are constructed, sustained and transformed. Similarly to the functionalist paradigm, it assumes a level of social order and regulation exists. Metaphors within this paradigm include Accomplishments, Enacted sense-making, Language games and Text.
- *Radical humanist paradigm*: This paradigm also takes a subjective view of reality, similar to the interpretive paradigm, but emphasises how this reality and aspects of society such as work, language and technology constrain and dominate humans and limit them from achieving their true potential. A single metaphor included in this paradigm is the Psychic prison.
- *Radical structuralist paradigm*: This paradigm posits an objective reality, which is understood in terms of the structures of society and how they dominate human activity. The focus is on understanding how conflicts and tensions within these structures eventually lead to radical change. The metaphors included in this paradigm are Instruments of domination, Schismatic systems and Catastrophe.

- *Enacted sense-making metaphor*: How do language and communication support sense-making?
- *Psychic prison metaphor*: Do actors feel manipulated by the system or by those exercising power in the system?
- *Schismatic systems metaphor*: Where are points of tension or conflict within the system?

The two sets of prompts were used by the research team in a group process to identify additional insights not generated during the analysis process. The full set of

CWA prompts was applied; however, due to time constraints, not all organisational metaphor prompts could be considered. Instead, four metaphors were randomly selected: *accomplishments*, *enacted sense-making*, *theatre* and *instruments of domination*. A sample of the insights generated through this process is given in Table 6.2.

TABLE 6.2
Examples of Insights Identified through the Prompting Process

Prompt Question	Insight	Insight Type
CWA Prompts		
Overall context: What is the system's most obvious weakness?	In terms of efficiency and time delays, there is no consistency. Road users do not know how long they will be waiting at active crossings.	Pain point
Control Task Analysis/contextual activity template: Are there situations involving high workload?	In the urban context, drivers experience high workload on approach to rail level crossings as they must monitor and avoid other road users (e.g. pedestrians, cyclists, other drivers) as well as process visual clutter in a dense built environment. To negate this, and draw attention to the crossing, a raised platform could be used in conjunction with a clearway before the crossing.	Design solution
Social Organisation and Cooperation Analysis: Would any tasks completed by humans be better completed by technology?	How can the vehicle be used to provide warnings to the driver?	Leverage point
Organisational Metaphor Prompts		
Accomplishments: What are the social rules or patterns that assist actors to successfully interact within the system?	How can we change social norms for non-compliant groups or reduce the possibility that the rail level crossing environment is used by individuals to challenge authority?	Leverage point
Enacted sense-making: Does the system support sense-making?	How can the system identify an obstruction on the tracks when the train is approaching and provide assistance?	Leverage point
Instruments of domination: Do processes or aspects of the system dominate or control actors within the system?	Pedestrians who want to catch an approaching train are punished for compliance (and may be rewarded for non-compliance by catching their intended train).	Pain point

6.3.3 Insight Prioritisation

A total of 209 insights were generated, including those documented spontaneously during the analysis process and those generated through applying the prompt questions. A prioritisation process was undertaken by the research team, which involved each member identifying those insights they thought were most likely to effectively contribute to the development of new designs for improving safety at rail level crossings. The more highly prioritised insights were then used to develop the materials for the participatory design workshop activities.

6.3.4 Design Process Planning

The CWA-DT provides templates for developing a design brief and documenting the design criteria. The design brief ensures that the design process is appropriate for the project and that it remains driven by the sociotechnical systems theory philosophy throughout any design application. It also enables clear communication about the background and purpose of the design process, which assists if additional stakeholders need to be briefed as the design process proceeds. It provides a short summary of the approach and may be supplemented with additional project management tools, such as detailed project schedules and budgets, depending on the size and scope of the project.

The design brief developed during the design planning stage of this project outlined the aim of the design task as 'to develop design concepts that will increase safety at Victorian public rail level crossings'.

The scope was constrained to improve at-grade interfaces, rather than the development of grade separation options (i.e. the construction of bridges or tunnels). Further, the focus was on shaping desired behaviour, rather than improving technological reliability. Finally, the designs were not intended to focus on reducing incidents involving intentional self-harm at rail level crossings; however, it was noted that it would be beneficial if design concepts introduced some positive indirect effects on such behaviour.

Key measures for determining the success of the design process should be documented in the design criteria template. It is good practice to define up front the criteria that will be used to measure project success. The design criteria should be drawn from the values and priority measures identified in the Work Domain Analysis phase of the CWA. They should also assess the extent to which design outcomes reflect sociotechnical systems theory principles.

In the current CWA-DT application, the design criteria were drawn from the Work Domain Analysis and from sociotechnical systems theory principles. In terms of the Work Domain Analysis criteria, the values and priority measures were selected with emphasis on the safety-related values: *Minimise* risk, collisions, trauma, injuries, fatalities, near-miss events and road rule violations; and *Maximise* efficiency, reliability and conformity with standards and regulations. It was determined that the designs should also align with the sociotechnical systems theory content principles to be successful.

In addition, the design criteria noted that a form of cost–benefit analysis would be required, given the limited government funding available to upgrade rail level crossings.

6.3.5 Design Tool Selection

Following the development of the design brief and specification of the design criteria, the design tool selection matrix from the CWA-DT was used to select the most appropriate tools and activities to be used in the workshop. The tools selected were as follows:

- *Sociotechnical values cards* to introduce sociotechnical systems theory values and promote value-aligned thinking
- *Personas* to communicate findings of the research and promote empathy for different system users
- *The impossible challenge exercise* to promote thinking outside of usual constraints (e.g. time, budget)
- *Scenarios* to communicate the findings of the research and provide an understanding of context
- *Assumption crushing* to promote lateral thinking and prompt design ideas
- *Metaphorical design* to promote lateral thinking and prompt design ideas
- *Inspiration cards* to prompt design ideas

6.3.6 Idea Generation Workshop

Eighteen participants attended a 2-day idea generation workshop. As can be seen in Figure 6.4, participants were representatives of rail level crossing stakeholder organisations (e.g. government departments, safety regulators, road user advocacy bodies, transport investigators) and those with a professional interest in the research (e.g. human factors professionals, researchers, designers). Although we did not have any attendees from the Government/Parliament level of the system, we expected that relevant perspectives for this level would be represented by the policy makers who attended from government departments.

During the workshop, design participants engaged in a range of exercises, using the tools selected by the research team during the design planning stage (see Section 6.3.5). Participants usually worked in small groups of four to five individuals.

Before beginning the structured exercises, participants undertook general brainstorming as an initial activity. This provided an opportunity for participants to express ideas that they may have already generated outside of the design process that were not influenced by the CWA findings or the sociotechnical systems theory philosophy. Participants were asked 'What are your ideas for improving rail level crossing safety?' They were asked to brainstorm and record their ideas individually on paper. They were then asked to share their ideas with the wider group. Although not expressed to participants, an intention of this warm-up activity was for participants to express any long-held beliefs around how to improve rail level crossing safety, so that they would be able to think about new ways to tackle the problem during the remainder of the workshop.

6.3.6.1 Sociotechnical Values Cards

This activity provided participants with information about the sociotechnical systems theory values that underlie the approach of the CWA-DT. To ensure participants

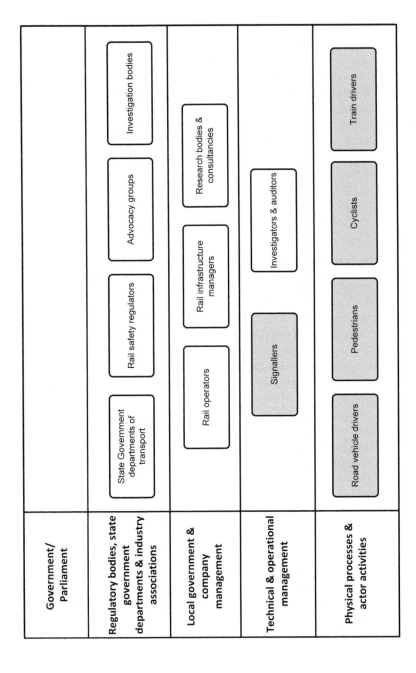

FIGURE 6.4 Attendees at design workshop, mapped onto Rasmussen's (1997) Actormap. Note that the actors shaded in grey represent secondary roles (i.e. most attendees were also the members of a road user group, two attendees were employed as signallers in previous roles and one attendee had previously been employed as a train driver).

understood and engaged with the values and philosophy, they were asked to discuss the values in small groups. To promote discussions around the values, sociotechnical values cards were created, with each card having the title of an sociotechnical systems theory value (see Chapter 3) and a picture to represent its meaning. For example, the card for 'Humans as assets' contained a picture of a ballerina to emphasise the point that humans have creative and artistic abilities that cannot be matched or replaced by technology.

The sociotechnical values cards were randomly distributed among the small groups so that each group received two values on which to focus. The groups were asked to consider the following questions for their values and document their discussion:

- What is your understanding of this value?
- Do you believe it is important?
- In what ways is the value supported at rail level crossings?
- In what ways is the value not supported at rail level crossings?
- How is this value relevant to today's workshop?

Groups then shared the key aspects of their discussion with the wider group to promote a shared understanding of how the value was relevant.

6.3.6.2 Personas

Personas (described to participants as user profiles) were used to encourage the development of empathy for different road user types, to enable them to better consider the user's perspective and to understand their motivations and challenges. They were based on generic characteristics of rail level crossing users. They included a name, occupation, road user type (pedestrian, cyclist, motorcyclist, car driver or heavy vehicle driver), main reason for travel and a short description of the person's travel patterns, personal preferences and motivations. The user profiles were intended to describe a concrete user, but were not intended to be overly stereotypical.

The handout included a space for participants to draw a sketch of the user as well as questions for participants to answer. The questions were intended to encourage participants to put themselves 'in the shoes' of the users. They included questions regarding the user's concerns when travelling generally, the user's concerns at rail level crossings, what the user likes about level crossings and the challenges the user has faced at level crossings. Participants were asked to consider each written profile and answer the questions, discussing this within their small group and reporting back to the larger group.

6.3.6.3 The Impossible Challenge Exercise

This exercise posed a problem for participants to solve, which had such tight constraints (e.g. time, budget, resources) that it could not be solved through rational, linear thinking (Imber 2009). Introducing such an exercise forces the brain to think laterally and acts as a 'warm-up' activity to promote lateral thinking during the remainder of the session. In the session, participants were posed an impossible challenge unrelated to rail level crossings (e.g. 'to play in the World Cup final') and worked in small groups to generate three solutions within 5 minutes.

6.3.6.4   Scenarios

For each rail level crossing context, participants were presented with scenarios that highlighted relevant research findings relating to that context, including important features of the rail level crossing functioning, as well as pain points and scenario features identified within the analysis insights. Each scenario focussed on a specific user type and, like the user profiles, described a fictional but typical, concrete experience. For example, for the context of urban rail level crossing adjacent to a railway station, the scenario involved a pedestrian running late for a train (see Box 6.4). The scenario handout included spaces in the text for participants to complete. These prompted the participants to consider the emotions of the road user to further promote empathy and understanding.

BOX 6.4   PEDESTRIAN SCENARIO

Dave is heading to work on a Tuesday morning. He has a board meeting at 9:00 am and he is thinking about the presentation he needs to give to outline the new marketing campaign; he is hoping to get up and running. He is hoping to get a seat on the train so he can make finishing touches to the PowerPoint presentation. Dave usually arrives at the train station early as he likes to be able to check his smartcard balance and not be rushed. However, this morning Dave's wife was feeling unwell so he dropped his daughter at swimming and brought the car back home, before hurrying with his briefcase and laptop bag towards the train station.

As Dave is approaching the station, he can hear the bells going for the Frankston train and knows he will have only a couple of minutes before the city-bound service arrives. He quickens his pace further and hopes his train has been delayed. The bells stop and he turns the corner onto McKinnon Road. Traffic that had banked up for the passing of the previous train has just begun to move. He takes the opportunity to cross over the road back from the crossing where the cars are still stopped. He makes eye contact with the driver he passes in front of to ensure they do not move their car forward before he is clear.

He runs towards to pedestrian crossing and reaches the fencing area just as the bells begin to sound again. He considers taking the underpass but thinks it will be quicker to cross the tracks, and anyway, the gates have not begun to shut yet. There is another pedestrian on the crossing who is talking on a mobile phone, and Dave has to move right over to the left to avoid colliding with them, while also pulling his wallet from his pocket to get out his smartcard. As he gets to the other side of the crossing, the gate is beginning to shut. He manoeuvres around it, runs up the ramp, swipes his smartcard and jumps on the train. He looks around and there are no seats left. Dave feels _____.

6.3.6.5   Assumption Crushing

Assumptions include theories, beliefs or hypotheses underpinning the structure of the system and the way things are currently done. They may not be consciously realised but can unconsciously restrict the type of design ideas that are considered. Assumption crushing provides a means to identify an alternative theory, belief or hypothesis, and brainstorm design ideas in line with this (Imber 2009). In the design process, participants were presented with assumptions identified during the insight generation process and were asked to develop alternative statements and then to identify design ideas based on those alternative assumptions.

The assumptions presented to participants were as follows:

- Road users are the cause of rail level crossing crashes.
- Decisions should be taken away from humans.
- We cannot fix safety without impacting on efficiency.

Participants were presented with the assumptions and were asked to undertake the following activities within their small groups:

- Discuss the extent to which you agree with the assumption.
- Brainstorm alternative statements.
- Choose an alternative statement and discuss how would you design a rail level crossing that aligns with this assumption.
- Document design ideas (using A5-sized sheets of coloured paper).

Following this activity, groups shared the key aspects of their discussion with the wider group.

6.3.6.6   Metaphorical Design

The use of metaphors and analogies is a common approach to design. These tools can bring new perspectives and prompt new ideas and innovation. The metaphor insights used with design participants were as follows:

- 'Separation' introduced with synonyms, including disconnection, detachment, severance, uncoupling, disjunction, segregation, division, gulf and chasm
- 'Barriers' introduced with the synonyms of fence, railing, barricade, hurdle, bar, blockade, roadblock, fencing, obstacle, obstruction, impediment, hindrance, deterrent, complication, difficulty, baulk and curb.

Participants were presented with each metaphor and were asked to undertake the following activities within their small groups:

- Brainstorm how separation is achieved or barriers are used in other areas/domains.
- Discuss how these lessons from other domains could be used at rail level crossings.
- Document design ideas (using A5-sized sheets of coloured paper).

Following this, participants shared the aspects of their discussion with the wider group.

6.3.6.7 Inspiration Cards

This activity involved the presentation of preselected inspiration cards, relevant to the design problem, to help prompt brainstorming in small groups. Participants were provided with a set of inspiration cards and were asked to use the cards to generate ideas for rail level crossing design.

The inspiration cards included the following:

- *Pain point cards*: Prioritised pain point insights identified from the analysis were printed on palm-sized, red cards (see Box 6.5 and Figure 6.5).
- *Leverage point cards*: Prioritised leverage points identified from the analysis were printed on palm-sized, pink cards (see Box 6.6 and Figure 6.5).
- *Cards from the Design with Intent toolkit*: Selected Design with Intent cards (Lockton et al. 2010) were used (see Figure 6.5). These cards present design patterns drawn from one of eight lenses or perspectives, including the architectural lens, ludic (playful) lens and cognitive lens. They provoke the consideration of opportunities or ideas from various design disciplines to influence behaviour.

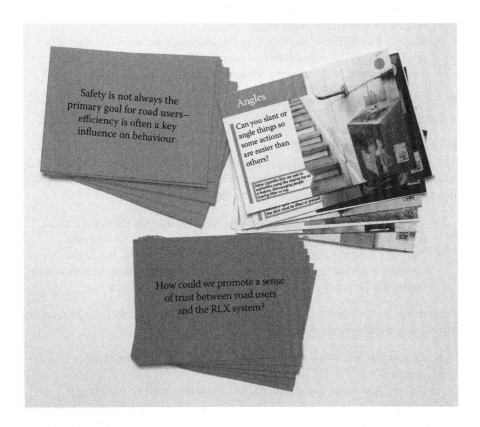

FIGURE 6.5 Inspiration cards used in the design workshop, showing pain point cards (top left), design with intent cards (top right) and leverage point cards (bottom). RLX, rail level crossing.

BOX 6.5   EXAMPLE PAIN POINTS USED
IN THE INSPIRATION CARD EXERCISE

- Humans are not good at judging the speed and distance of large objects.
- People are reluctant to stop the flow of movement.
- Safety is not always the primary goal for road users – efficiency is often a key influence on behaviour.
- No information is provided about train approach at passive crossings.
- Rural rail level crossings are typically found on high-speed limit roads.
- Expectancy plays a key role in user behaviour at rail level crossings, for example, if a user is not expecting a train, he/she may not check for it and may not perceive warnings.
- Urban environments are complex with many road user movements, visual clutter and reduced sight distance for trains.
- Users are not provided with information about whether another train is coming.
- Congestion can cause users (on road and pedestrian crossings) to become caught on the crossing when a train is approaching.
- Cyclists often do not have separate facilities for crossing the tracks.

BOX 6.6   EXAMPLE LEVERAGE POINTS USED
IN THE INSPIRATION CARD EXERCISE

- How can road users be engaged and empowered to take responsibility for their own safety at crossings?
- How could we provide the sense of a realistic and imminent threat at the crossing? How can we increase the sense of danger to prompt safe behaviour?
- How can we minimise the negative experience of waiting for the train? Can we make waiting time appear to go faster?
- How can we remind users of their underlying values around safety in a way that will encourage them to select that goal over others (e.g. efficiency)?
- How could we change the rail system to reduce the amount of waiting time for road users (e.g. through train stopping patterns)?
- How could we provide train drivers with control or a sense of control over what is happening at the crossing?
- How could we use people's preference to comply with social norms to influence their behaviour?
- How could vehicles be used to provide information to road users or improve safety at rail level crossings?
- How can we trigger automatic or habitual desirable behaviours?
- How can we communicate actual levels of risk dynamically?

Participants were given the flexibility to choose the way in which the cards were used and were asked to document any design ideas inspired by the cards on the A5-sized sheets of coloured paper provided in the session.

6.3.6.8 Impossible Challenge Revisited

Towards the end of the idea generation activities, we returned to the impossible challenge exercise – this time posing a challenge relevant to rail level crossing design. During this exercise, each group of participants was allocated a different existing rail level crossing context (urban or rural) and given the following challenge:

> The Queen is coming to visit Victoria next week and the premier has announced that she will open a new, world-leading rail level crossing upgrade. You have a budget of $1,000. What will you do in 1 week to the existing crossing to make it world-leading?

Participants were asked to generate and document their ideas to meet this challenge on the A5-sized sheets of coloured paper provided.

6.3.6.9 Design Concept Definition

Following the idea generation activities, participants were asked to create more comprehensive design concepts. To assist this, the ideas generated during the previous activities were displayed around the room (across the tables and walls). There were more than 150 design ideas generated (see Figure 6.6 for examples).

In addition to the participant-generated design ideas, additional ideas suggested by the researchers (i.e. design solution insights identified during the CWA) and ideas proposed in previous research were displayed around the room. Participants worked

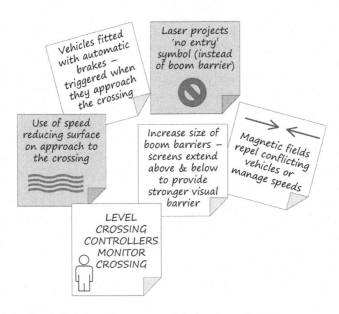

FIGURE 6.6 Example design ideas generated during the workshop.

in small groups, browsed through the ideas available and used one or a combination of ideas as inspiration for creating a more holistic design concept.

To document the design concepts, participants used templates that provided space to describe the concept using words, diagrams and/or sketches (see Figure 6.7). The large sheets incorporated prompts to give the concept a name, identify the context for the concept (metro, rural or both), provide a drawing or sketch of the design, describe the expected benefits, identify which road users the design would be effective for and identify the sociotechnical systems theory values addressed by the design. These latter aspects were intended to ensure that participants were reminded of the need to consider all road users and the sociotechnical systems theory values. Eleven design concepts were identified through this process and each was presented to the larger group.

6.3.6.10 Design Concept Prioritisation

Following the presentation of the concepts, participants were asked to consider each of the design concepts developed and to prioritise them in relation to how effective they were likely to be in improving safety at rail level crossings. Participants were asked to determine their own process for prioritising designs. They chose to use a voting system whereby each participant could indicate their first, second and third preferences, which were tallied to derive a ranked order of the 11 design concepts.

The top ranked five concepts were prioritised for further evaluation and refinement. They are as follows:

- *Comprehensive Risk Control crossing (for urban environments)*: This concept uses a combination of safety risk controls and amenity improvements. It focusses on alerting road users to the presence of the rail level crossing and approaching trains (through addition of traffic lights, advanced warning signs, in-road studs, default closed gate for pedestrians). It also aims to avoid queuing on the crossing or to mitigate its consequences (e.g. using traffic light coordination, 'hold' or 'keep tracks clear' sign for approaching traffic, awareness campaigns, an emergency lane and a no standing zone on the exit lane from the crossing), as well as to enforce rules (e.g. through camera enforcement, channelised fencing for pedestrians) and to provide convenience and amenity to waiting pedestrians (e.g. with an all-cross pedestrian phase, a shelter, community hub, ticketing machine and cafe at or near the waiting area).
- *Intelligent Level Crossing (for urban and rural environments)*: This concept is based around the use of new and emerging technologies to optimise transport system functioning by improving communication and coordination between the road and the rail. It provides decision support systems to road users (via in-vehicle interfaces, smartphones or dynamic displays), reduced delays and enforcement of stopping when collisions are predicted.
- *Heavy Vehicle Average Speed to avoid a train (for rural environments)*: This concept was intended for use by professional heavy vehicle drivers and involved using new and emerging technologies to provide drivers with

Design concept name: _____

Context (rural, metro, both): _____

Description (text & sketch):

Expected benefits:

_____

_____

Other system changes required:

_____

_____

Evaluation considerations:

_____

_____

Potentially effective for:
□ Drivers
□ Heavy vehicle drivers
□ Pedestrians
□ Motorcyclists
□ Cyclists

Values addressed:
□ Humans as assets
□ Technology as a tool to assist humans
□ Promote quality of life
□ Respect for individual differences
□ Responsibility to all stakeholders

FIGURE 6.7 Design concept template.

speed guidance that would enable them to avoid needing to stop for an approaching train.
- *Simple But Strong (for rural environments)*: The philosophy behind this design was to use simple and low-cost features to draw attention to the upcoming rail level crossing and the danger posed, and to provide warning of a train approach. It included a low-cost active warning sign (potentially solar powered), as well as use of colour (e.g. orange, red) on the 'danger zone' over the crossing and on signage at the rail level crossing.
- *Speed, Expectancy, Gap (for rural environments)*: This concept employed graduated speed reductions and rumble strips to slow road vehicles and provide time to look for trains as well as to provide opportunities to recover from errors. Similar to the Simple But Strong concept, this concept also employed a low-cost active warning sign to warn of train approach. Finally, a red light on the front of the train would be used to increase train conspicuity and draw attention to the hazard in the environment.

6.4   SUMMARY

This chapter has described the process undertaken to generate initial design concepts for urban and rural rail level crossings. It has demonstrated how the CWA-DT can be used to translate the outcomes of CWA and other analysis methods into a participatory design process with system stakeholders to create novel design concepts.

The design workshop was well received by participants, with high levels of engagement in the process among the diverse participants represented. The activities appeared to facilitate creative thinking and draw out important discussions about practicality and feasibility of different approaches.

Although the process generated five design concepts for further development, the ideas were generally quite high level and would require further work. This refinement process involved the evaluation of these initial concepts through the systems models developed in Phase 2, as described in Chapter 7.

7 Initial Design Concept Evaluation

With Contributions from:
Christine Mulvihill and Nicholas Stevens

7.1 INTRODUCTION

This chapter outlines the process for undertaking a desktop evaluation of design concepts generated using the Cognitive Work Analysis (CWA) Design Toolkit, which was described in Chapter 6. It shows how models of an existing system generated via CWA and Hierarchical Task Analysis (HTA) can be used in a systematic and robust way to model how system functioning and behaviour may be positively or negatively impacted by the introduction of new components. This evaluation can inform decision-making regarding which initial designs should proceed to further detailed design efforts and enables design teams to consider whether potential unanticipated negative consequences can be addressed through design refinement activities. This form of desktop design evaluation is important as it provides a low-cost approach for testing initial design concepts in a manner that is congruent with the analyses upon which the design concepts were built.

The five highest ranked design concepts from the idea generation workshop were evaluated. Concepts were selected based on the prioritisation process undertaken by stakeholders, as well as the research teams' judgements regarding alignment of each design concept with systems thinking, practicality and likelihood of implementation.

The desktop evaluation process involved the following four main activities:

1. Insertion of concepts into the existing rail level crossing Work Domain Analysis (WDA; see Chapter 5)
2. Analysis of potential user errors using HTA and Systematic Human Error Reduction and Prediction Approach (SHERPA; see Chapter 5)
3. Evaluation of each design concept against sociotechnical systems content principles (see Chapter 6)
4. Evaluation of each design concept against the key risks to rail level crossing safety identified in Chapter 5

Each activity is demonstrated through examples of how it was achieved for one of the design concepts we developed and evaluated. The different evaluation approaches enabled the consideration of the extent to which the designs aligned with systems thinking, together with the likely impact and emergent behaviours associated with each design concept.

7.2 DESIGN EVALUATION WITH CWA

CWA is uniquely suited to the evaluation of the impacts of system change due to its formative nature. That is, it enables the analyst to identify all of the possible effects on system functioning, not just those that might be expected or anticipated. It provides a structured thinking tool that can help to identify emergent behaviours, unintended consequences of change, potential risks introduced due to system changes and potential design requirements to address these.

In this research programme, evaluation with CWA was conducted using large hardcopy diagrams of the appropriate WDA model (i.e. passive or active rail level crossing) as a starting point for the evaluation. Evaluations were conducted by small teams, with one analyst documenting the changes to the WDA diagram and another documenting the process and resulting discussion using an electronic template. For two concepts, the team included a small number of stakeholders who participated in the design workshops, to familiarise stakeholders with the process to enable future applications to proposed design changes without assistance from the researchers. Due to the time required to conduct each evaluation, the involvement of stakeholder representatives in the evaluation process was not feasible for all five concepts and other evaluations were completed by the research team only.

To undertake the WDA evaluation, the following steps were undertaken (see Figure 7.2 for an example):

- New physical objects present in the design concept were added to the bottom level of the WDA. For example, if the design concept included an in-vehicle warning device, then this was added to the physical object level of the WDA.
- Existing physical objects enhanced in the proposed design were highlighted. For example, an improved rail level crossing warning sign would enhance the existing physical object 'rail level crossing signage'.
- For each new and enhanced object, the following was undertaken:
  - The existing object-related processes that it would support were highlighted. For example, the new object 'In-vehicle warning device' would enhance the object-related processes 'Alert user to presence of rail level crossing' and 'Alert user to presence of train'.
  - Pathways following the means-ends links from these highlighted object-related processes up to the functional purposes of the system were highlighted.
  - Any new object-related processes that the new or enhanced object would afford were added into the abstraction hierarchy.
  - Pathways following the means-ends links from these new object-related processes up to the functional purposes of the system were highlighted.

As the changes were being overlaid onto the WDA, a standardised template was used to document the following information arising from discussions:

- Assumptions being made about the design and the effect it will have on system functioning.
- Key benefits that are apparent from reviewing the effects on the system.
- Potential negative effects, such as new risks introduced by the new design.
- Areas where further investigation or research is required to understand the impacts.
- Suggested refinements to improve the design's potential to achieve desired benefits or to minimise potential negative effects of the new design.
- The frequencies of new nodes, nodes that are enhanced or supported, nodes that are appropriately restricted and nodes that are negatively influenced.

The final step in the WDA evaluation process was to assess the impact of each new object assessed by summing the following at each level of the abstraction hierarchy:

- *New nodes*: For example, the new physical object 'Optimal speed to avoid train in-vehicle display' would 'Communicate optimal speed' and 'Provide distance to rail level crossing' notification.
- *Support for existing nodes*: For example, the new physical object 'In-vehicle warning display' would provide support for the existing function of 'Alert user to the presence of train'.
- *Appropriate restriction*: For example, the new physical object 'Default closed pedestrian gates' would appropriately restrict (pedestrian) traffic flow, which in turn would support the function of 'Maintain road and rail user separation'.
- *Negative influence*: For example, the new physical object 'Speed limit reduction signs' would have the effect of slowing traffic through rail level crossings, which in turn may negatively influence the 'Maximise efficiency' value and priority measure.

Each of the five prioritised design concepts was evaluated in this manner. To illustrate, we present the evaluation of the Speed, Expectancy, Gap concept.

7.2.1 Evaluation of Speed, Expectancy, Gap Concept with WDA

The Speed, Expectancy, Gap design concept was intended for rural road environments and proposed the use of low-cost features to draw attention to the upcoming rail level crossing and warn of an approaching train. The technologies were intended to be low cost in comparison with existing rail level crossing warning devices. Low-cost devices might use alternative train detection technologies, power sources (e.g. solar) and wireless communications technologies (Wullems et al. 2013). In addition, road user speed would be managed on approach, through graduated reductions in the posted speed limit, to provide more time to recover from errors.

The design included the following features:

- Gradual reduction of speed limit for drivers on approach (from 80 km/h to 60 km/h, then 40 km/h).
- The train would trigger a loop in the track (or GPS could be used) to activate signs facing the road and a high-intensity flashing red light on the front of the train.
- Signs at the crossing incorporate a rail level crossing symbol, the text 'give way to trains' and twin flashing lights that activate when a train is approaching. The sign has a 'healthy-state' light for train drivers to monitor whether the technology is functioning. One sign would be placed on approach to the rail level crossing (as an advance warning), with another located at the rail level crossing itself.
- Road markings and rumble strips would be present on approach to and at the crossing – red curved lines with text reading 'RAIL X'.

The original design concept template from the idea generation workshop is shown in Figure 7.1.

Evaluating this concept with WDA involved taking the diagram of the existing passive rail level crossing system and overlaying the changes that would occur if this design was implemented. The passive rail level crossing WDA was selected because it was intended that this design would be used to 'upgrade' an existing passive rail level crossing, rather than replacing an active crossing in a rural environment. An extract of the WDA with the changes is shown in Figure 7.2.

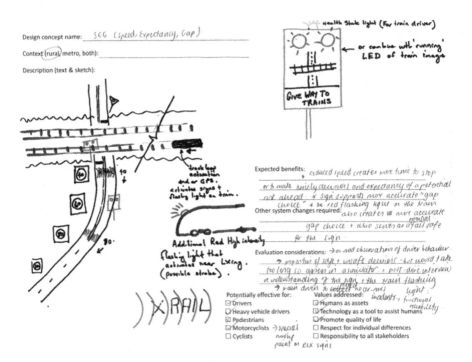

FIGURE 7.1   Speed, Expectancy, Gap initial design concept.

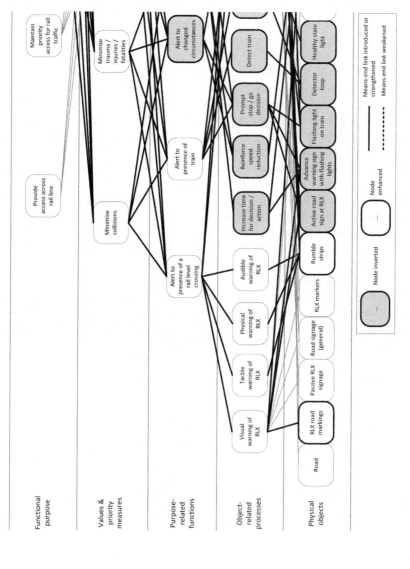

FIGURE 7.2 Extract of WDA evaluation of the Speed, Expectancy, Gap concept. Note that only the left side of the WDA is shown due to space restrictions. RLX, rail level crossing.

While the WDA evaluation was being undertaken, the findings were documented in the template developed for this activity (see Table 7.1 for an example of a completed template). Of particular interest were the potential errors or risks introduced. Two potential risks were associated with the speed limit reduction, which could mean that vehicles (especially heavy vehicles) have increased exposure to the risk of being struck by a train as they would require longer to traverse the crossing. The other was an introduced violation: it was thought that road users would be unlikely to comply with lowered speed limits, particularly when no train was approaching. Rural environments in which the design would be used typically have speed limits of 100 km/h, which reduce to 80 km/h approximately 400 m prior to the rail level crossing and 100 m after the crossing. Therefore, the reduction to 40 km/h is very slow in comparison. Those involved in the evaluation process felt that there was some benefit in retaining the low speed, as it could be expected that road users will slow somewhat in response; however, introducing rules that are unlikely to be complied with was a concern as it could undermine compliance in other areas of the road system.

Another potential risk identified through this evaluation related to the issue of providing information to road users, which may then lead them to speed up, to try to cross prior to the train arriving at the rail level crossing. This relates to the earlier decision ladder analyses that suggested that road users were more likely to engage in non-compliant behaviour when they have access to additional information about the system state (see Chapter 5). In the Speed, Expectancy, Gap design concept, the additional information was in the form of the flashing light on the train and the advance warning sign. This may be of higher concern in locations where long, slow freight trains are common and road users are frustrated by having to wait for extended periods of time. Interestingly, there is evidence that drawing attention to the front of the train may in fact support more accurate judgements of train speed and thus support safer decisions (Clark et al. 2016).

7.3   DESIGN EVALUATION WITH HTA AND SHERPA

Identification of potential user errors for each design concept was achieved by reviewing the HTA and SHERPA outputs (described in Chapter 5) and using them to predict the likely errors that could arise when users interact with the new design. The process for evaluating the concepts through HTA and SHERPA involved the following:

- Reviewing the HTA and updating it with:
  - New tasks
  - Changes to existing tasks
- Using SHERPA to identify potential errors associated with the new and changed tasks
- Reviewing the entire SHERPA analysis, considering the following:
  - Are existing errors removed?
  - Do new tasks provide recovery opportunities for existing errors?
  - Is the probability or criticality of errors affected by the changes?

TABLE 7.1

Findings from WDA Evaluation of Speed, Expectancy, Gap Concept

Assumptions	Effects on WDA	New Risks/Errors Introduced	Further Investigation	Suggested Improvements
• Rumble strips are appropriate for the speed limit (e.g. can be traversed at 80 km/h). • The flashing light on train would only activate when the train is approaching an RLX. • Detector loop is more reliable than GPS (therefore assumed that detector loop would be utilised). • Lowering speeds on approach would give road users more time to check for trains.	**Physical objects** *New objects (total = 5)* Active road sign at RLX Advance warning sign with lights (flashing) Flashing light on train Detector loop Healthy state light *Enhanced objects (total = 3)* RLX road markings Rumble strips Speed limit signs **Object-related processes** *New processes (total = 6)* Increase time for decision/action Reinforce speed reduction Prompt stop/go decision Detect train Trigger crossing warning Fault detection *Existing processes supported (total = 11)* Visual warning of RLX Tactile warning of RLX Audible warning of RLX	• Reducing speed could result in trucks/vehicles on the rail level crossing for longer – increases the risk of being struck. • Could introduce more road rule violations as road users may not comply with lower speed restrictions (although it is expected that they will slow down somewhat more than if limits were not present). • Some users may see the light on the train or the advanced warning sign and speed up to cross before the train arrives.	• Research undertaken on the effectiveness of strobes on trains found they were ineffective, need to check the scope of research. • Is there research on the effectiveness of text-based road markings? • Investigate rumble strip design that is appropriate for higher speeds for all vehicle types (including two-wheelers).	• Provide a variable message sign that gives feedback to drivers if they are exceeding the speed limit on approach. • Include a picture of a train on speed limit signs to reinforce they are linked to a rail level crossing – to create expectation that arrangement of signs means approaching an RLX. • Education campaign to inform about the rationale for introducing speed reductions. • Train light could be activated by GPS and the sign activated by the detector loop to provide redundancy. • Active warnings should have a failure mode that displays when the technology is not functioning. Needs to be understood by road users, could be linked to healthy-state light/indication for train drivers.

(Continued)

TABLE 7.1 (*Continued*)
Findings from WDA Evaluation of Speed, Expectancy, Gap Concept

Assumptions	Effects on WDA	New Risks/Errors Introduced	Further Investigation	Suggested Improvements
• Assume that the speed reduction signs are rolled out consistently across regional areas so that road users will expect them at RLXs. • That if road vehicles, especially trucks, slow more gradually on approach, there will be less impact on the road surface and subsequently less maintenance required.	Visual warning of approaching train Attract attention Cue attention Communicate road rules Discomfort Speed reduction Communicate requirement to stop/give way Disseminate information on risk and appropriate behaviour *Processes negatively influenced (total = 0)* **Functions** *Existing functions supported (total = 5)* Alert to presence of RLX Alert to presence of train Behave appropriately for environment Maintain road user and train separation Maintain infrastructure *New functions (total = 1)* Alert to changed circumstances *Functions negatively influenced (total = 0)*			• Detector loop could optimise the warning time to ensure consistent warning time is experienced by road users.

(Continued)

TABLE 7.1 (*Continued*)
Findings from WDA Evaluation of Speed, Expectancy, Gap Concept

Assumptions	Effects on WDA	New Risks/Errors Introduced	Further Investigation	Suggested Improvements
	Values and Priority Measures (V&Ps)			
	Existing V&Ps supported (total = 5)			
	Minimise collisions			
	Minimise trauma/injuries/fatalities			
	Minimise risk			
	Minimise road rule violations			
	Maximise reliability			
	V&Ps negatively influenced (total = 0)			
	Discussed the impact on the efficiency of slowing road vehicles but decided it was negligible.			
	Functional purposes			
	Existing functional purposes supported (total = 4)			
	Protect road users			
	Protect rail traffic			
	Minimise delays to rail traffic			
	Minimise delays to road network.			
	Purposes negatively influenced (total = 0)			

RLX, rail level crossing

The output was a series of likely errors for each concept, including a description of each error and the associated consequences, ratings of probability (low, medium, high) and criticality (low, medium, high) and potential remedial measures. Metrics such as number of existing potential errors reduced by the new design concept as well as new errors introduced were also calculated.

7.3.1 EVALUATION OF SPEED, EXPECTANCY, GAP CONCEPT WITH HTA AND SHERPA

HTA and SHERPA were used to evaluate the Speed, Expectancy, Gap concept. On reviewing the HTA 21 operations were added, with responsible actors shown in parentheses:

- Look for speed limit signage (road user)
- Detect speed limit signage (road user)
- Interpret speed limit signage (road user)
- Look for active signage at rail level crossing (road user)
- Detect flashing lights at rail level crossing (road user)
- Interpret flashing lights at rail level crossing (road user)
- Activate active advance warning signage (train and rail level crossing)
- Activate flashing light on train (train)
- Activate active road sign at rail level crossing (train and rail level crossing)
- Look for active advance warning sign (road user)
- Detect flashing lights on advance warning sign (road user)
- Interpret flashing lights on advance warning sign (road user)
- Look for speed limit signage (road user)
- Detect speed limit signage (road user)
- Interpret speed limit signage (road user)
- Look for flashing light on train (road user)
- Detect flashing light on train (road user)
- Interpret flashing light on train (road user)
- Deactivate active early warning signage (rail level crossing)
- Deactivate active warning signage at rail level crossing (rail level crossing)
- Deactivate flashing light on train (train)

SHERPA was then applied to identify the potential errors relating to each new operation. Existing errors were also reviewed to determine whether the new design would eliminate any existing errors or provide recovery opportunities. An extract of the SHERPA analysis showing examples of the new errors introduced is provided in Table 7.2, and an extract of the analysis of existing errors is shown in Table 7.3.

TABLE 7.2

Extract of SHERPA Evaluation Showing New Errors Identified

Task	Error Mode	Error Description	Consequence	Recovery	P	C	Remedial Measures
Activate flashing light on the front of the train (train)	A8 – Operation omitted	Train fails to activate flashing light on the front of the train	Train fails to announce the presence of train to road users	• Activate flashing light signage at RLX (train)	M		Back-up detectors
Activate flashing light on the front of the train (train)	A2 – Operation mistimed	Train activates flashing light on the front of the train too late	Train fails to announce the presence of train to road users	• Activate flashing light signage at RLX (train)	M		Back-up detectors
Activate signage at RLX (train and RLX)	A8 – Operation omitted	Train fails to activate signage at RLX	Train fails to announce the presence of train to road users	• Activate flashing light on the front of the train (train)	M	✓	Back-up detectors
Look for flashing light on train (driver)	C1 – Check omitted	Road user fails to see flashing light on train	Road user fails to comprehend approaching train	• See train (driver) • Sound horn (train driver)	M	✓	In-vehicle technology
Detect flashing light on train (driver)	R1 – Information not obtained	Road user fails to detect flashing light on train	Road user fails to comprehend approaching train	• See train (driver) • Sound horn (train driver)	M	✓	In-vehicle technology
Detect flashing light on train (driver)	C5 – Check mistimed	Road user detects flashing light too late	Road user fails to comprehend approaching train	• See train (driver) • Sound horn (train driver)	M	✓	In-vehicle technology

C, criticality; H, high; L, low; M, medium; P, probability; RLX, rail level crossing.

TABLE 7.3

Extract of SHERPA Evaluation Showing Changes to Existing Errors Identified

Task	Error Mode	Error Description	Consequence	Recovery	P	C	Remedial Measures
Detect presence of train	R1 – Information not obtained	Road user fails to detect approaching train	Road user fails to comprehend train is approaching and does not stop	~~None~~ • *Activate flashing light on the front of the train (train)* • *Activate flashing light signage at RLX (train and RLX)*	~~M~~ L	✓	In-vehicle technology
Sound horn (train driver)	A8 – Operation omitted	Train driver fails to sound train horn	Train fails to announce the presence of train to road users	~~None~~ • *Activate flashing light on the front of the train (train)* • *Activate flashing light signage at RLX (train and RLX)*	~~M~~ L	✓	In-cab reminder system

Note: Aspects of the existing system analysis that are no longer relevant are struck out; additions are shown in italics.
C, criticality; H, high; L, low; M, medium; P, probability; RLX, rail level crossing.

7.4 DESIGN EVALUATION AGAINST SOCIOTECHNICAL SYSTEMS THEORY PRINCIPLES

Following evaluation using the CWA outputs, the designs were also evaluated against the sociotechnical systems theory content principles, as outlined in the design criteria document. A description of each of the content principles can be found in Chapter 6.

To conduct the evaluation, a set of indicators was developed to assist the analyst to judge whether the principle is 'present' in the proposed design (see Table 7.4). The evaluation process considers the design concept as a whole, so the analyst must judge the extent to which the whole concept aligns with the indicators provided. This process provides some basis for discriminating between different concepts in terms of their overall alignment with STS and an indication of whether design concepts align with individual principles strongly or weakly. Design concepts meeting all indicators for a principle receive a score of 3, whereas those meeting none of the indicators receive a score of 1. Concepts that partially meet the criteria are given a rating between 1 and 3, with half ratings permitted. With 14 content principles, the highest rating possible is 42 and the lowest is 14.

This evaluation process is best conducted in a group setting to increase the chance that no aspects of the design concept are overlooked. Further, this process can prompt refinements to the design concept which are documented to inform design refinement processes.

An excerpt of the STS evaluation for the Speed, Expectancy, Gap concept is shown in Table 7.4. Overall, the ratings of the concept against each of the criteria were quite low. This suggests that this design concept was not well aligned with sociotechnical systems theory philosophy.

7.5 KEY RISKS ADDRESSED

As described in Chapter 5, a list of key risks was identified from the original CWA, HTA and SHERPA analyses. The research team considered each of the key risks and determined whether the design concepts would address the risk. The findings for the Speed, Expectancy, Gap design concept are shown in Table 7.5.

7.6 SUMMARISING THE EVALUATION RESULTS FOR EACH DESIGN CONCEPT

A Concept Evaluation Summary template (adapted from Liedtka and Ogilvie 2010) was used to provide a structured means for documenting the results of the evaluation. This template facilitates the documentation of the key needs addressed by the design concept, the approach taken in the design concept (i.e. what are the overall changes and what philosophical underpinnings are associated with the change), the key benefits of the proposed design, the estimated costs of the changes, potential negative effects such as new risks or potential for human error and the assumptions made during the evaluation process. This summary is intended to support

TABLE 7.4

Extract of the STS Evaluation of the Speed, Expectancy, Gap Design Concept

Content Principle	Indicators/Measures	Score	Comments	Potential Refinements
Tasks are allocated appropriately between and among humans and technology	• Users are provided with appropriate tasks, that is, not monotonous tasks that can be performed by technology with human supervision, but those that provide appropriate challenge. • Users are given tasks where there is unpredictability such that judgement and interpretation are needed.	2	• Users retain the task of choosing when to stop, but are given additional warnings. • Users are slowed in stages to give the feeling of preparing to stop. • Train drivers are given a healthy-state light to monitor. However, there is little they can do immediately to assist the situation if the warnings have ceased to work beyond using the train horn to alert road users and reporting the issue.	• The red light should activate at the point where there would be a collision if the user chooses to cross. This ensures the credibility of the warning (e.g. it is not activated for 200 m prior to the RLX if the train is travelling at 20 km/h). • Road users should be given a failure mode display for the sign. • Road users should be educated about the meaning of the sign and the flashing light on the train.
Useful, meaningful and whole tasks are designed	• Users are given whole tasks, rather than fragmented pieces of tasks to perform. • Users perceive the tasks they are performing to be meaningful or useful.	1	• Users would understand the significance of the task, but there is nothing added to give the task meaningfulness. • Speed reductions and rumble strips after the RLX will be perceived as unnecessary.	• Speed reductions and rumble strips should be present only on approach – the limit should lift to 80 km/h, then 100 km/h shortly after the RLX.
Boundary locations are appropriate	• Boundaries or divisions between users or user groups are based on whole processes. • Boundaries do not impede sharing of knowledge and experience.	1.5	• The design introduces a new form of communication between the train and the road user via the light on the train. This helps to communicate across boundaries. • Nothing included in the design to share knowledge and experience. • User groups not considered include rail maintainers who operate track vehicles.	• May need to install red lights on non-train vehicles that use the railway (e.g. track machines). Will there be means for them to engage a test switch to start the warnings if not detected by the train detection loop?

(Continued)

TABLE 7.4 (Continued)
Extract of the STS Evaluation of the Speed, Expectancy, Gap Design Concept

Content Principle	Indicators/Measures	Score	Comments	Potential Refinements
Boundaries are managed	• The design facilitates supervisors, users or user groups to manage the boundaries or interfaces with other users or groups. • Boundary management incorporates buffering from external disturbances or changes in the wider environment. • The design empowers users to define the rules and processes that constrain their activities.	1	• No boundary management is present. Instead, the design relies on road rules being followed. • The flashing light sign would be advisory rather than regulatory – meaning that people can to some extent choose to ignore it in certain circumstances. • Users are unable to define rules and processes.	• Potential to introduce a feedback system online so users can give feedback about the timing of sign and train light activation.
Problems are at their source	• The design facilitates the detection of and recovery from problems (including negative behaviours) at the time and place at which they occur. • The design provides people with the competency and authority to control problems.	1.5	• Train drivers are given a healthy-state light to monitor. However, there is little they can do immediately to assist the situation if the warnings have ceased to work beyond using the train horn to alert road users and reporting the issue. • Road users are not given a failure mode display for the low-cost signs. • The graduated speed reduction means users will have opportunity to see another sign if one is missed (currently just one sign for 80 km/h limit is present).	• Provide a failure mode for the sign (e.g. a sign displaying 'no warnings provided – look for trains'). • Could use portable speed feedback signs on approach to the RLX.

RLX, rail level crossing.

TABLE 7.5

Evaluation of the Speed, Expectancy, Gap Concept and Its Impact on Reducing Key Risks

Risk	Addressed by Design Concept?
Road users not aware of upcoming rail level crossing	Yes – through rumble strips, enhancement of road markings, speed signs on approach and rail level crossing warning signs
Road users not aware of rail level crossing warnings	Yes – through the addition of advance warning signs and warning signs at the rail level crossing with flashing lights
Road users not aware of an approaching train	Yes – through the addition of a red light at the front of the train to draw attention, as well as the addition of advance warning signs and warning signs at the rail level crossing with flashing lights
Road user not checking for trains sufficiently	Yes – the speed reductions are intended to provide road users with more time to check for trains. Additionally, the red light on the train is intended to attract attention to an oncoming train
Road user not detecting a second or subsequent train	The design does not specifically address this risk
Road user misjudging the speed or distance of the train	Yes – the red light on the front of the train is intended to draw attention to the train and assist road users to make decisions about its speed and distance from the rail level crossing
Road user choosing to cross when warnings are activated/a train is approaching	The design does not specifically address this risk
Road user queuing or short stacking on the crossing	The design does not specifically address this risk
Warning systems failing to announce the presence of a train	This risk may be increased in this design, as a low-cost warning device is proposed

discussions with design stakeholders about selecting a particular design or shortlist of designs and the need for design refinements prior to moving into the detailed design stage.

7.7 COMPARING DESIGNS

In addition to summarising the findings for each design concept, metrics were calculated based on the design activities to enable comparison of the design concepts, as shown in Table 7.6.

In terms of the WDA measures supported by each of the designs, the Heavy Vehicle GPS design best supported the values and priority measures, with the objects introduced in this design supporting all but one. However, a metric was also identified based on a change-to-benefit ratio, which divided the number of nodes supported with the WDA by the number of new objects added. This metric highlights designs

TABLE 7.6
Summary of Design Concept Evaluation Findings

Evaluation Element	Comprehensive Risk Control Crossing	Intelligent Level Crossing	Speed, Expectancy, Gap	Simple But Strong	Heavy Vehicle GPS
WDA – Values and Priority Measures Supported					
Minimise collisions	Y	Y	Y	Y	Y
Minimise trauma, injuries and fatalities	Y	Y	Y	Y	Y
Minimise near-miss events	Y	Y	N	N	Y
Minimise risk	Y	Y	Y	Y	Y
Maximise efficiency	N	Y	N	Y	Y
Minimise road rule violations	Y	N	Y	Y	Y
Maximise reliability	N	N	Y	N	N
Maximise conformity with standards and regulations	N	N	N	Y	Y
Total (*maximum possible 8*)	5	5	5	6	7
WDA – Change-to-Benefit Ratio					
Total number of supported nodes divided by the number of added nodes (larger number better)	2.7	4.4	2.3	3.7	1.7
Road Users Considered					
Drivers	Y	Y	Y	Y	N
Heavy vehicle drivers	Y	Y	Y	Y	Y
Pedestrians	Y	Y	N	N	N
Motorcyclists	Y	N	Y	Y	N
Cyclists	Y	N	N	Y	N
Total (*maximum possible 5*)	5	3	3	4	1

(*Continued*)

TABLE 7.6 (Continued)
Summary of Design Concept Evaluation Findings

Evaluation Element	Comprehensive Risk Control Crossing	Intelligent Level Crossing	Speed, Expectancy, Gap	Simple But Strong	Heavy Vehicle GPS
Key Risks Addressed					
Road users not aware of upcoming RLX	Y	N	Y	Y	Y
Road users not aware of RLX warnings	Y	Y	Y	Y	Y
Road users not aware of an approaching train	N	Y	Y	Y	Y
Road user not checking for trains sufficiently	N	N	Y	N	N
Road user not detecting a second or subsequent train	N	Y	N	N	N
Road user misjudging the speed or distance of the train	N	N	Y	N	Y
Road user choosing to cross when warnings are activated/a train is approaching	Y	N	N	N	N
Road user queuing or short stacking on the crossing	Y	Y	N	N	N
Warning systems failing to announce the presence of a train	N	Y	N	N	Y
Total	4	5	5	3	5
Human Error Analysis					
Total number of errors possible	138	104	113	83	93
Error potential (number of introduced errors divided by number of components) (smaller number better)	6	6.1	7.1	9.2	9.3
Average number errors per concept eliminated/added (compared to existing active crossing)	−1.1	−1.0	0.0	2.1	2.2
Percentage of total errors where recovery supported	59.4%	53.8%	68.1%	51.8%	57.0%

(Continued)

TABLE 7.6 (*Continued*)
Summary of Design Concept Evaluation Findings

Evaluation Element	Comprehensive Risk Control Crossing	Intelligent Level Crossing	Speed, Expectancy, Gap	Simple But Strong	Heavy Vehicle GPS
Sociotechnical Systems Theory Principle Alignment Scores					
Combined score of evaluation against the 14 sociotechnical systems theory principles (total score between 14 and 42, larger number better)	18.5	20	20.5	19.5	19.5
Cost					
	VH (single RLX)	VH (multiple RLXs)	VH (single RLX, all trains)	M-H (single RLX)	H (multiple RLXs)

Very high (VH) = >$500,000
High (H) = $300,000–$500,000
Medium (M) = $100,000–$300,000
Low (L) = <$100,000

RLX, rail level crossing.
Y = Yes.
N = No.

for which a small amount of change can have large positive impacts across the system. Here, the Speed, Expectancy, Gap concept and the Comprehensive Risk Control crossing appear to provide the most benefit.

When considering the next aspect of evaluation, the road user types considered, the limitations of the Heavy Vehicle GPS concept become apparent with it only being targeted at one road user group. This means the benefits may be more limited than other concepts such as the Comprehensive Risk Control crossing, which incorporates design features for all user types. In addition, none of the rural design concepts considered pedestrian users, however, this was a deliberate omission as these designs were intended for high speed roads with minimal to no pedestrian traffic.

In relation to the key risks, all the designs addressed the risk of failing to notice the rail level crossing warnings. This suggests that the design process may have resulted in concepts that extended the status quo, rather than proposing revolutionary change in the design philosophies applied. Specifically, the historical focus underpinning rail level crossing design has been on warning the road user of the train's presence and requiring the road user to give way to the train.

Apart from this, the evaluation summary clearly demonstrates where risks were only addressed by a single design. For example, only the Speed, Expectancy, Gap concept addressed the risk of users not checking for trains sufficiently (i.e. by lowering the approach speed), and only the Comprehensive Risk Control crossing specifically addressed the instances of road users crossing when the warnings were activated or a train was approaching (i.e. road rule violations).

A comparison of the findings from the human error analysis conducted with HTA and SHERPA provided some additional interesting findings. Although some concepts, such as the Comprehensive Risk Control crossing and the Intelligent Level Crossing, had a relatively high frequency of possible errors, their error potential ratio was lower than for the other concepts. Furthermore, this analysis highlighted that most of the designs did not cater for error recovery particularly well, although the Speed, Expectancy, Gap concept was the most promising in this regard, because of the slower speed limits introduced.

In relation to the evaluation against the sociotechnical systems theory principles, the designs generally did not fully align with this philosophy. The concepts all scored lower than half of the possible top score of 42. This may suggest that although the design process assisted in moving towards more sociotechnical systems theory-based designs, it was not fully successful in achieving this aim. It may be that the design process led participants to focus on physical interactions within the rail level crossing system, preventing sociotechnical systems theory philosophies from being fully realised. It is likely that rail level crossing design concepts that fully align with sociotechnical systems theory would be radically different from existing rail level crossing environments. However, radical designs can be difficult to achieve within an existing domain where there are significant practical obstacles, including the cost of changes and the influence of existing institutional and legal structures.

Finally, the research team discussed with stakeholders the likely cost estimates of the new designs to better understand the likely practical limitations on implementation. Here, we see that although the project was attempting to develop low-cost solutions, due to the inclusion of multiple components within each design concept the costs were generally rated as very high (e.g. above the average cost of upgrading a passive crossing to an active crossing). For example, in the Speed, Expectancy, Gap concept, the implementation of the red flashing light on all trains across multiple train operating companies was estimated to be very high cost, whereas the warning signs could be implemented at specific locations at low cost.

It should be noted that for some designs the costs need not be borne by road or rail authorities. For example, concepts based on in-vehicle assistive devices would likely be part of larger efforts to introduce such technologies into the road transport system. Thus, other actors such as vehicle manufacturers and vehicle consumers (including heavy vehicle companies) may bear some of the implementation costs. Other costs, such as the installation of shelters for pedestrians at the Comprehensive Risk Control crossing, could be offset by using the infrastructure for commercial advertising.

7.8 SUMMARY

This chapter has described how design concepts can be evaluated in a desktop manner to identify the potential benefits and risks associated with their implementation. It demonstrates how the systems-based models, originally developed to understand the existing system and generate insights for its re-design, can be re-purposed to understand the effects of change.

Although the evaluation process provided some interesting findings into the potential effectiveness of the different concepts, the findings were not always clear-cut. It was initially expected that the evaluation process would provide a well-defined ranking of the most effective to least effective design concepts. Instead, it highlighted the positive and negative aspects of the designs and, importantly, helped the research team to identify the important refinements to the designs.

The process of undertaking the evaluations was quite lengthy, given the detail in the initial models and the systematic process undertaken. This made it difficult to involve the entire research team in all evaluation activities and limited the extent to which the evaluation could be conducted in a participatory manner with stakeholders. Future work could consider whether evaluations can be limited to consideration of key design aspects or key risks. This could particularly be useful for the HTA and SHERPA evaluations; potentially, evaluation could focus on impacts to only critical tasks and critical errors. The process could also be made more efficient by extending the software tools used to develop the initial models to support the evaluation process.

Although the process was lengthy, it still provided a cost-effective means to conduct a high-level evaluation, prior to further design refinement activity and detailed design. It also ensured that design concepts were evaluated in a manner consistent with the analyses that informed the design concepts in the first place. This is a critical requirement that is not often achieved. The approach is novel in the rail level crossing field: anecdotally, it appears that proposed designs are assessed on a cost and engineering practicality basis and, if promising, will be proposed for further testing and trials in driving simulators or in the field. This is likely because there have not been accessible human factors evaluation methods available to practitioners working in this area. It is hoped that this intermediate step of desktop evaluation provides additional tools that can be useful to understand the potential risks and benefits of proposed designs. In Chapter 8, we describe how the findings from the desktop evaluation process were used to refine the design concepts.

8 Design Concept Refinement

With Contributions from:
Christine Mulvihill, Nicholas Stevens and Guy Walker

8.1 INTRODUCTION

This chapter describes how we used the findings from the desktop evaluation process to refine the novel rail level crossing (RLX) design concepts. This involved a two-step process, which incorporated first a participatory process of design refinement with stakeholders, followed by an expert process. In addition, this chapter describes an expert process that was undertaken to generate an additional two design concepts. Finally, the six RLX designs that were selected for detailed evaluation and testing with prospective users are described in detail.

8.2 STAKEHOLDER DESIGN REFINEMENT WORKSHOP

The design concepts generated using the Cognitive Work Analysis Design Toolkit (CWA-DT) were refined through a participatory workshop involving rail level crossing stakeholders. The design refinement workshop was held approximately 6 months after the initial idea generation workshop described in Chapter 6. This enabled the desktop evaluation activities (see Chapter 7) to occur in the intervening time frame.

Ten participants, all representatives from road or rail stakeholder organisations, attended the workshop. Eight participants had previously attended the idea generation workshop. Prior to the design refinement workshop, participants were provided with a written summary of the evaluation of the five novel design concepts, including the comparison between concepts (see Table 7.6). The summary document included the findings from the desktop evaluation using the Work Domain Analysis (WDA), the Systematic Human Error Reduction and Prediction Approach and the sociotechnical systems theory principles. For each design concept, the summary outlined the components incorporated in the design, the key risks addressed, the potential negative effects, costs and suggestions for improvement identified during the evaluation process. An example of the concept evaluation summary for the Speed, Expectancy, Gap concept is presented in Table 8.1.

At the beginning of the workshop, attendees were given an introductory presentation that re-introduced participants to systems thinking and the sociotechnical systems theory design philosophy, and the design concepts, as well as describing the findings from the desktop evaluation process.

TABLE 8.1
Concept Evaluation Summary Provided to Participants for the Speed, Expectancy, Gap Concept

Concept Summary: Speed, Expectancy, Gap (Rural)

Approach	This concept incorporates a gradual speed limit reduction to build an expectation that there is something coming up which requires response and to build the expectation to stop (speed limit signs and rumble strips). The rumble strips also contain the Rail-X message to alert road users that the reason for speed reduction is an RLX ahead. The design also incorporates low-cost warnings to alert road users of train approach (active advance warning sign and sign at the RLX, red flashing light on the train).
Components
• Gradual reduction of speed for drivers on approach (100, 80, 60 and 40 km/h).
• Train activates loop (or GPS used) to activate signs facing road and flashing red light on the front of the train.
• Light on train is an additional red high-intensity flashing light that activates near the crossing (possibly strobe).
• Signs at the RLX with symbol of RLX, text 'give way to trains' and twin flashing lights (active when train approaching). Sign has healthy-state light for train drivers to monitor. Alternative image for sign could be 'running' LED of train image. One sign provided on approach (as advance warning), and the other is located at the RLX.
• Road markings and rumble strips on approach and at the crossing – red curved lines with text 'X RAIL'.
Key needs addressed (from WDA)	Minimise collisions ✓
Minimise trauma, injuries, fatalities ✓
Minimise near-miss events
Minimise risk ✓
Maximise efficiency
Minimise road rule violations ✓
Maximise reliability ✓
Maximise conformity with standards and regulations

(Continued)

TABLE 8.1 (*Continued*)
Concept Evaluation Summary Provided to Participants for the Speed, Expectancy, Gap Concept

Concept Summary: Speed, Expectancy, Gap (Rural)

Benefits	*Behaviours/precursors addressed or mitigated*
	• Road users not aware of approaching RLX (e.g. through looked but failed to see errors, expectancy issues)
	• Road users not aware of RLX warnings (e.g. through looked but failed to see errors, expectancy issues)
	• Road users not aware of approaching train (e.g. through looked but failed to see errors, expectancy issues, misinterprets warnings)
	• Road user does not check for trains sufficiently
	• Road user misjudges speed/distance of train
	Contributing/escalating factors addressed or mitigated
	• Monotonous environment (e.g. rural roads)
	• Expectancy
	Additional benefits
	• Flashing red light on the train is particularly useful for heavy vehicle drivers that need to sight train from longer distance to be able to cross RLX safely.
	• Slowing vehicles on approach to the RLX would reduce wear and tear on the road surface – increase reliability and reduce costs.
Potential negative effects	• Could introduce more road rule violations with the speed limit restrictions – road users may not comply with lower restrictions.
	• Some road users may see the light on the train or the advanced warning sign and speed up to cross before the train.
	• Reducing the speed could mean that trucks/vehicles are on the RLX for longer increasing the risk of being struck (need to take into account in determining the warning cycle).
	• Rumble strips may be an issue for two-wheelers and heavy vehicles.
Further investigation required	• Research on effectiveness of strobes on trains found was ineffective – but scope of research may be limited.
	• Need to consider the impacts on other train drivers of using the red flashing light to warn road users – will be seen by other train drivers who may interpret it as a requirement to stop. Consider other colours.
	• Investigate effectiveness of writing on the road surface.
	• Investigate rumble strip design that is appropriate for higher speeds for all vehicle types (including two-wheelers).

(Continued)

TABLE 8.1 (*Continued*)
Concept Evaluation Summary Provided to Participants for the Speed, Expectancy, Gap Concept

Concept Summary: Speed, Expectancy, Gap (Rural)

Suggested improvements	• Warning sign could have a graphic of a train approaching in the direction from which they are actually approaching (including subsequent trains where appropriate). This would assist to inform road users that the lights activate when a train approaches.
	• Consider providing a warning phase with the sign, for example, orange lights flash, then red lights flash when train is closer.
	• Provide a variable message sign that gives feedback to drivers if they are exceeding the speed limit on approach, also if they have traversed the RLX late/experienced a near miss. The sign could also communicate information about recent near misses and incidents. Messages are used to promote responsibility for own and others' safety.
	• Include a picture of a train on speed limit signs to reinforce they are linked to an RLX – to create expectation that arrangement of signs means approaching an RLX.
	• Ensure speed limit sign placement is appropriate – road users are not being slowed for a longer period than necessary.
	• Have speed reductions and rumble strips only on the approach to the RLX (not the exit) to retain credibility. On exit, the limit should raise to 80 km/h, then 100 km/h shortly after the RLX.
	• Public education campaign to inform about the intention of introducing the speed reduction signage and to educate on the operation of the new sign and the flashing light on the train.
	• The red flashing light on the train could be activated by GPS and the sign at the RLX activated by the detector loop to provide redundancy so if one fails another should still operate.
	• The red flashing light on the train could be linked to the train horn so it is activated for a certain period after the horn is sounded for the RLX – cheaper and potentially useful for other rail warnings.
	• The flashing light on the train could only activate when there could be a collision if the road user crosses, to retain warning credibility. Factors in train speed, not just its location.

(Continued)

TABLE 8.1 (*Continued*)
Concept Evaluation Summary Provided to Participants for the Speed, Expectancy, Gap Concept

Concept Summary: Speed, Expectancy, Gap (Rural)

- Active warnings should have a failure mode that displays when the technology is not functioning (similar to traffic lights flashing yellow). Must be understood by road users, could be linked to healthy-state light/indication for train drivers. For example, the sign could state – 'no warnings provided – look for trains'.
- The train driver receives a notification if the red flashing light is not working.
- The detector loop could optimise warning time to ensure consistent warning time is experienced by road users.
- Design rumble strips to be acceptable to heavy vehicle drivers.
- Provide a shoulder/lane for cyclists and motorcyclists that does not include rumble strips – maintains a smooth surface as much as possible.
- Provide the red flashing light on track machines and road-rail vehicles or provide a means for them to engage a test switch to activate the warnings if they are not detected by the train detection loop.
- Provide an online feedback system so users can give feedback about timing of sign activation and train red flashing light activation.
- Local implementation of the design could involve making it more visually appealing – associated with the local context such as local industry and environmental features.

RLX, rail level crossing; GPS, global positioning system; LED, light-emitting diode.

8.2.1 Design Improvement Review

Following familiarisation with the concepts, participants worked in groups of four to five, with each group facilitated by a member of the research team. A booklet including the design refinement suggestions for each concept was used to present each suggestion and record the consensus of the group (i.e. accept suggestion, reject suggestion) and the reasons for these decisions. Table 8.2 shows a sample of the design refinements from the Speed, Expectancy, Gap concept, and the feedback and decisions of the workshop groups. Additional changes to concepts proposed by the groups were also recorded. Each group presented their agreed refinements to the broader group for discussion.

8.2.2 Evaluation and Ranking of Concepts

Once the concepts were refined, a large, printed scoreboard displaying the evaluation criteria was used for a group discussion to compare the refined concepts. The scoreboard displayed the following criteria:

- Whether the design concept supported the values and priority measures from the WDA (e.g. minimise collisions, maximise efficiency, minimise road rule violations)
- Whether the design concept took into account the needs of different road user types (e.g. drivers, cyclists, pedestrians)
- Whether the design concept addressed or mitigated key risks associated with rail level crossings (e.g. road user not aware of an approaching train, road user queues or short stacks on the crossing)
- The estimated cost of implementing the proposed designs (high, medium or low)
- A rating of the level of innovation demonstrated by the design concept (a rating from 1 to 5, with 5 being the highest level of innovation)

The workshop participants discussed each of the criteria for each design concept and provided a consensus rating or ranking on each. At the conclusion of this process, each participant was provided with three voting tokens, which they were asked to use to vote for the design concepts that they felt best met the evaluation criteria and should be prioritised for further detailed design and testing processes. Voting was achieved by placing the tokens at the bottom of the scoreboard aligned with the chosen design concept. The concepts receiving the highest votes were the two intelligent transport system (ITS) concepts: the Intelligent Level Crossing and the Heavy Vehicle GPS concept (renamed the GPS Average Speed concept following a decision during these discussions to extend the concept to all drivers), as well as the Comprehensive Risk Control crossing. Following group discussion, it was decided that the Speed, Expectancy, Gap and Simple But Strong concepts (both of which individually received relatively low votes) had appreciable similarities, which lent themselves to be combined for further evaluation; the combined design was thereafter referred to as the Simple But Strong crossing.

TABLE 8.2

Sample of Proposed Refinements and Feedback on the Speed, Expectancy, Gap Concept

Proposed Refinement	Purpose/Benefits	Positive Comments	Negative Comments	Decision
Warning sign could have a graphic of a train approaching in the direction from which they are actually approaching (including subsequent trains where appropriate). This would assist to inform road users that the lights activate when a train approaches.	• Provides additional credibility to the warning – users understand the risk and why they are being asked to stop. • Clarity that the lights only flash when a train is approaching. • Avoid second train coming incidents.	n/a	• Would only work in a single-line context. • Additional information may overload drivers. • Needs a fail-safe display – would be technologically complex and costly.	Reject
Consider providing a warning phase with the sign, for example, orange lights flash, then red lights flash when train is closer.	• Provide more information about the risk and credibility to the stop message (i.e. it is never safe to be on the RLX when the red lights are flashing). • Reduce the likelihood of panic or vehicle-to-vehicle collisions caused by sudden braking.	n/a	• Would not work on the advance warning sign (if a red light is given, this indicates need to stop). • Technologically complex, similar to upgrade to full active crossing (flashing lights and boom barriers) – cost implications.	Reject
Ensure speed limit sign placement is appropriate – road users are not being slowed for a longer period than necessary.	• Provide credibility to the speed limit signage – more likely to comply with limits if it seems reasonable. • Reduce road user frustration.	• Seen as obviously beneficial	n/a	Accept
The red flashing light on the train could be linked to the train horn so it is activated for a certain period after the horn is sounded for the RLX – would be more economical and potentially useful for other rail warnings.	• More economical way to implement the idea – links in with existing technology fitted on the train. • Provide train drivers with control over the flashing light – they can use it more or less depending on the circumstances (e.g. if stopped at the train station near the RLX it will not flash continuously). • Additional benefits – for example, may draw attention of trespassers on the track.	n/a		Accept

(Continued)

TABLE 8.2 (*Continued*)
Sample of Proposed Refinements and Feedback on the Speed, Expectancy, Gap Concept

Proposed Refinement	Purpose/Benefits	Positive Comments	Negative Comments	Decision
The flashing light on the train could only activate when there could be a collision if the road user crosses, to retain warning credibility. Factors in train speed, not just its location.	• Provide an additional credibility to the warning – only flashes when it is dangerous to cross.	n/a	• Considered too complicated to implement.	Reject
Active warnings have a failure mode that displays when the technology is not functioning (similar to traffic lights flashing yellow). Must be understood by road users, could be linked to healthy-state light/indication for train drivers. For example, the sign could say – 'no warnings provided – look for trains'.	• Failures of the technology to warn of a train are indicated to users so they know to look for trains rather than relying on the signal.	n/a	• Considered too costly to implement, propose to use a passive stop sign as the failure mode.	Reject
Provide the red flashing light on track machines and road rail vehicles or provide a means for them to engage a test switch to activate the warnings if they are not detected by the train detection loop.	• Ensure road users get warnings of any rail vehicles approaching the RLX, not only trains.	n/a	• Considered unnecessary as vehicles other than trains are required to stop prior to traversing RLXs.	Reject

n/a, not applicable; RLX, rail level crossing.

8.3   DESIGN PROCESS EVALUATION

8.3.1   Participant Reflections on the Participatory Design Process

At the conclusion of both the initial design workshop (discussed in Chapter 6) and the design refinement workshop, attendees completed evaluation questionnaires to enable us to gain their perceptions of the process.

The questionnaires administered after the initial design workshop asked about a range of attributes of the design process, rated on a 5-point scale from 'strongly disagree' to 'strongly agree'. Overall, participants rated the process as meeting the attributes of facilitating creativity, providing structure to the design process, enabling holistic thinking, efficiency, enabling integration into existing design processes, facilitating iteration of ideas and demonstrating validity (Read et al. 2016a).

In relation to the overall process, considering both the initial idea generation workshop and the design refinement workshop, participants were asked whether they felt STS values had been followed within the process. The ratings received were generally very positive:

- 90% of participants agreed or strongly agreed that people were treated as assets.
- 80% agreed or strongly agreed that the technology was used as a tool to assist humans.
- 90% agreed or strongly agreed that the workshops promoted attendees' quality of life.
- 100% agreed or strongly agreed that individual differences were respected during the process.
- 90% agreed or strongly agreed that the workshops promoted consideration of responsibilities to all stakeholders.

In addition, both questionnaires asked open-ended questions about the positive and negative aspects of the process in which they had participated. The positive themes identified included the following:

- Collaboration (e.g. 'Design was fun in group setting', 'Open discussion, all views considered and valued')
- Creativity (e.g. 'Looking at effective solutions and provide alternative solutions', 'The creative approach was refreshing')
- Structure (e.g. 'It was well structured and relevant', 'Well structured and facilitated, good tools and prompts, especially Design with Intent cards and assumption crushing')

In terms of areas for improvement, the themes identified included the following:

- Limited time to achieve the workshop scope (e.g. 'We covered a lot of different elements... Perhaps more time – however appreciating the limited time available – it appears that a lot of information was still produced for consideration')

- A stronger focus on cost-effectiveness (e.g. 'Consideration of cost-effectiveness to enable informed discussion with the budget holders and support business cases/investment proposals', 'Looking for low cost solutions to ensure that we comply with the law')
- Ensuring the best mix of skills and expertise among workshop participants (e.g. 'A planned process of workshops so they are spread out over time so that we can optimise the people who can attend', 'Pre-qualification of participants – ability to work in teams, relevant knowledge, diversity of skills')

Together, these comments on the process suggest that the attendees appreciated the opportunity to work with other stakeholders in a collaborative manner and to think more creatively than they usually would in their everyday work. However, they wanted the process to be extended to incorporate cost–benefit considerations in more detail and suggested ways in which the process could be planned to ensure comprehensive consideration from those with different areas of expertise.

8.3.2 Researcher Reflections on the Participatory Design Process

As a research team, we perceived a high level of engagement from stakeholders in the initial design workshop and considered a number of the ideas that emerged to be innovative. However, discussions in the design refinement workshop emphasised the wider constraints on innovation in rail level crossing design. While reviewing the potential design refinements that were intended to improve alignment with socio-technical systems theory and systems thinking generally, attendee feedback was appreciably more conservative and the majority of the refinements were rejected.

The concerns raised fell generally around the following themes:

- *Standards and regulations*: The need to change standards to accommodate new designs was seen as a barrier, as were concerns about compliance with legislation. For example, the road rules would need to be updated to refer to new types of warning devices.
- *Government policy*: Some existing policies in road design were discussed in relation to a number of refinements, including the need to avoid distracting drivers with information not directly relevant to the immediate driving task, and ensuring that regulatory signs (e.g. speed signs) are not implemented where they will not be seen as credible (thus encouraging violations).
- *Cost*: Attendees were particularly concerned about the cost of potential designs given the importance of cost–benefit considerations for decision makers. They noted that cost–benefit analyses would need to be calculated to enable comparison between concepts.
- *Organisational, social and political considerations*: A range of other organisational, social and political issues were also raised, for example, how acceptable the designs would be to the wider community.

In general, refinements based on sociotechnical systems theory were seen as peripheral and unnecessary, suggesting that current design decisions are based on normative assumptions. That is, the focus appeared to be on tasks in isolation (i.e. 'can you see the rail level crossing sign without being distracted?') rather than more holistic considerations, such as ensuring road users understand how rail level crossing function and what functional purposes are trying to be achieved.

Feedback at the design refinement workshop also highlighted the tension between the aims of academic research and the needs of industry in partnering with research organisations. In the present research programme, the research team were focussed on developing new designs via a novel approach that was theoretically underpinned by sociotechnical systems theory and systems thinking. Conversely, for the project stakeholders, the goal was to develop options that can be practically implemented given the constraints under which they work. This meant that revolutionary elements of the design concepts often had to be modified to make them more practical. Although the participatory process intended to manage the balance between these goals and find solutions that could achieve both, in practice this was difficult to achieve. A key requirement for the human factors discipline generally lies in developing approaches that can effectively balance the goals and needs of research and industry.

8.4 RESEARCHER DESIGN REFINEMENT ACTIVITIES

Following the participatory design process, there was a need to generate more detailed design specifications for the in-vehicle assistive devices, to enable user testing using driving simulation. During the design refinement workshop, stakeholders accepted a proposed refinement to the Intelligent Level Crossing and the GPS Average Speed concept, which was to generate the detailed designs for the in-vehicle devices using the Ecological Interface Design (EID) approach. This was proposed by the researchers to provide an additional link between the initial systems analyses (i.e. CWA) and the design concepts.

In addition, given the assessment that the designs generated through the participatory process did not radically change rail level crossing (e.g. as reflected by poor alignment with STS principles), an additional design workshop was held with the research team to generate more revolutionary concepts.

8.4.1 In-Vehicle Interface Design Using EID Principles

EID is a design strategy that uses the abstraction hierarchy from WDA, coupled with principles from the skills, rule and knowledge taxonomy (Rasmussen 1983, Vicente and Rasmussen 1992).

In this taxonomy, skill-based behavior is the lowest level of cognitive control and refers to sensory-motor performance, which occurs in skilled activity without the requirement for conscious processing. Rule-based behaviour involves the application of stored rules, based on past experience, to make decisions. Finally, the highest level of cognitive control is knowledge-based behaviour, which is engaged during decision-making in unfamiliar situations where it is not possible to draw upon past experience. In these situations, reasoning is used to understand the situation and select an appropriate course of action.

The principles of EID specify that an interface should not require an operator to employ a higher level of cognitive control than necessary for the demands of the task. Further, the interface should support each level of cognitive control (i.e. skill, rule and knowledge-based behaviour). The underlying philosophy of EID is that the design (i.e. display or interface) should make the system constraints explicit to its end users. As with CWA, the focus is on the overall system and its constraints. As such, EID aims to make the interface transparent; its goal is to support direct perception and action, while correspondingly providing support for problem-solving activities (Vicente and Rasmussen 1990).

EID has been applied to the design of interfaces within varied domains ranging from nuclear process control (e.g. Burns et al. 2008) and health care (e.g. Watson and Sanderson 2007), to road transport (e.g. Young and Birrell 2012). Experimental evaluations have demonstrated that ecological interfaces elicit better performance than traditional interfaces (for reviews, see Burns and Hajdukiewicz [2004] and Vicente [2002]).

Although EID was not initially considered as a candidate approach for rail level crossing design due to our focus being the design of the physical environment and the infrastructure, it was incorporated into the process following the inclusion of in-vehicle interfaces in two proposed design concepts. The initially proposed designs for the in-vehicle interfaces provided warnings about approaching rail level crossing and approaching trains using symbolic representations (e.g. pictorial images of a warning sign shown on the interface). Such representations are perceived as signs, which activate rule-based processing (Rasmussen 1983). They are limited as they require the observer to understand conventions (e.g. by drawing on experience or training), and they cannot be used to look beyond what is presented to engage in troubleshooting or problem solving when there is a failure or other disturbance.

Separate detailed design workshops were undertaken to revise the designs of in-vehicle interfaces for the Intelligent Level Crossing and the GPS Average Speed concepts. In each workshop, the WDA was reviewed to identify the key constraints that should be displayed to the driver; then options for representing these constraints were generated and refined, and a final selection was made by the research team.

For the Intelligent Level Crossing concept, the key constraints identified were the train itself (its position, speed and direction of travel), the approaching vehicle (its position, speed and direction of travel) and the relationship in time and space between the two. It was determined that the display should dynamically represent the field of safe travel for the road user (Gibson and Crooks 1938). Field of safe travel theory posits that drivers operate by perceiving a dynamic space around their vehicle which, based on the surrounding traffic and hazards, is judged to be safe to occupy. Drivers seek to preserve an acceptable ratio between the available stopping distance and the boundary of the perceived safe field ahead. Drawing on this notion, we designed a visual representation of the dynamic field of safe travel, which would be presented ahead of the vehicle in the form of a green 'tongue' that shrinks in size and eventually disappears as the vehicle approaches the crossing while the train is also approaching (see Figure 8.1a for an early design sketch). This visual tracking would support skill-based behaviour, whereas additional features such as a symbolic representation of the train appearing on the interface to indicate an approaching train, and auditory messages stating the direction of train approach, would assist in supporting rule- and knowledge-based behaviour.

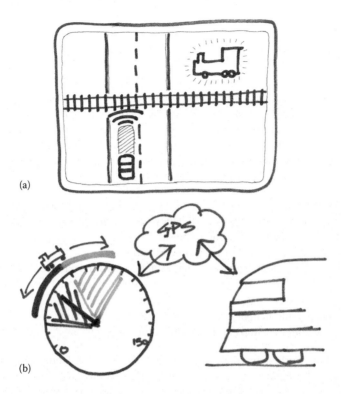

(a)

(b)

FIGURE 8.1 Design sketches for in-vehicle assistive devices using EID principles. (a) Early design sketch for the urban Intelligent Level Crossing in-vehicle interface and (b) Early design sketch of the GPS Average Speed in-vehicle interface.

For the GPS Average Speed concept, the key innovation was speed management of the vehicle as it approached the crossing, advising users to make small speed adjustments over a long distance so they could avoid coming to a complete stop for the train. Here, the key constraints were speed and position of the train and road vehicle, with the display providing guidance on the speed required for the driver to avoid arriving at the crossing at the same time as a train. In order to best support skill-based visual tracking, the design interface was overlaid onto the driver's speedometer (see Figure 8.1b for an early design sketch of the interface). This would provide the driver with an interval of desired or safe speeds (a green zone shown on the display), and a lower and upper speed limit, with the upper limit considered undesirable or unsafe (a red zone shown on the display). These zones would be dynamically updated by analysing the road vehicle's position and speed, compared with the train's position and speed. Again, to also support rule- and knowledge-based behaviour, a train symbol was shown on the display when the speed guidance was active, to convey that the train approach was the reason for the guidance. In addition, an auditory tone would be used to alert the user when the speed guidance becomes active and another tone would be sounded when the vehicle was travelling at a speed within the designated red zone.

8.4.2 GENERATION OF ADDITIONAL DESIGN CONCEPTS

In addition to using EID to develop specifications for the in-vehicle interfaces, the design team participated in a workshop to generate two additional design concepts. The workshop applied design tools from the CWA-DT, with the aim of developing more revolutionary rail level crossing designs to compare with the concepts generated via the participatory design process. Six members of the research team attended the expert design workshop. Attendees received the same design brief as given to those involved in the participatory process, with the instruction to generate one new rail level crossing design for an urban environment and one new design for a rural environment.

The activities undertaken included the following:

- *Assumption crushing*: The assumptions used were as follows: 'We can't fix safety without impacting on efficiency' and 'Rail level crossings have to have warnings at the crossing to be safe'.
- *Metaphors*: The metaphor used was 'Separation'.
- *Constraint crushing*: Attendees reviewed the active and passive WDAs, identified the key constraints and for each constraint considered the following:
  - What would happen if we removed the constraint?
  - What would happen if we strengthened the constraint?
  - How could we make hidden constraints visible?

The team then worked in two groups to use the insights from the activities to develop two additional concepts: the Community Courtyard crossing for urban environments (see Section 8.5.3) and the EID Crossing concept for rural environments (see Section 8.6.2).

8.5 FINAL DESIGN CONCEPTS FOR URBAN ENVIRONMENTS

After the final design workshop, there were a total of three concepts designed for use in an urban setting: two design concepts generated through participatory workshops with stakeholders (Comprehensive Risk Control crossing, Intelligent Level Crossing) and one design concept generated in the final workshop with the research team (Community Courtyard crossing).

8.5.1 COMPREHENSIVE RISK CONTROL CROSSING

The Comprehensive Risk Control crossing concept uses a combination of risk controls to separate road users from trains and from one another, as well as aspects to improve amenity, particularly for pedestrians. It focusses on drawing road users' attention to the presence of the rail level crossing and approaching trains (e.g. through the addition of traffic lights, advanced warning signs, in-road studs and a default closed gate for pedestrians). It also aims to avoid queuing on the rail level crossing and to mitigate its consequences (e.g. using traffic light coordination, 'hold' and 'keep tracks clear' signs for approaching traffic, an awareness campaign, emergency lane and no standing zone on rail level crossing exit), as well as to enforce rules (e.g. through camera enforcement and channelised fencing for pedestrians). Finally, it provides convenience and amenity to waiting pedestrians (e.g. through an

FIGURE 8.2   The Comprehensive Risk Control Crossing.

all-cross pedestrian phase, shelters, a community hub, ticketing machines and cafés at or near the waiting area).

Figure 8.2 shows a computerised sketch of the concept, whereas Figure 8.3 over-lays the components of this design on Rasmussen's (1997) framework, lines between components indicate where there are relationships between components. As seen in Figure 8.3, the concept incorporates a large number of additions to the physical environment of the crossing, as well as some interventions that operate at the higher levels of the system.

8.5.2   Intelligent Level Crossing

The Intelligent Level Crossing concept is based around the use of new and emerging intelligent transport systems (ITS) technologies to better optimise the functioning of the transport system by improving communication and coordination between road and rail systems. It provides decision support systems to road users (via in-vehicle assistive devices, smartphones or dynamic displays), reduced delays and enforce-ment of stopping when collisions are predicted. The components of the design con-cept are illustrated in Figure 8.4.

A key component in the concept is the in-vehicle assistive device, which was designed based on the EID philosophy described previously. It incorporates the fol-lowing features (as shown in Figure 8.1a):

- On approach to the rail level crossing when a train is approaching, the in-vehicle display provides an audible alert tone and a visual train icon appears on the display.
- A green 'tongue' appears on the display to indicate the safe field of travel on the road ahead.
- As the vehicle moves closer to the rail level crossing, curved bars appear (in line with the stop line) to show the limit of the field of safe travel.
- When no train is approaching, the display continues to show a representation of the roadway ahead but shows no indication (i.e. no green tongue is displayed).

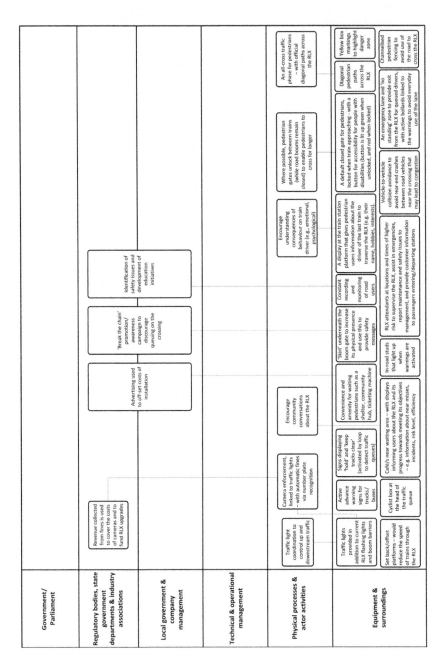

FIGURE 8.3 The Comprehensive Risk Control Crossing design concept overlaid onto Rasmussen's (1997) framework. RLX, rail level crossing.

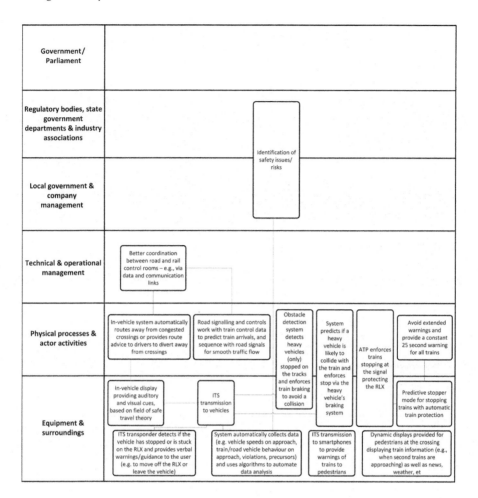

FIGURE 8.4 The Intelligent Level Crossing design concept displayed on Rasmussen's (1997) framework. ATP, automatic train protection; RLX, rail level crossing.

8.5.3 COMMUNITY COURTYARD CROSSING

The Community Courtyard crossing concept is underpinned by the notion of shared space and the prioritisation of active transport in the roadway. It also provides a vibrant space to enhance social interaction and inclusion while providing a focus of transit-orientated economic and community activity. It involves delineating the area around the rail level crossing and giving this area a courtyard atmosphere. Traffic is held back away from the crossing using traffic lights, and the road is raised in the courtyard space with no clear separation between the footpath and the roadway. The courtyard itself is a shared space where motorised traffic is expected to give way to non-motorised users, particularly pedestrians. To facilitate this, the road speed limit is reduced in this area, as is the speed of the train. Finally, standard rail level crossing warning devices

FIGURE 8.5   The Community Courtyard design concept.

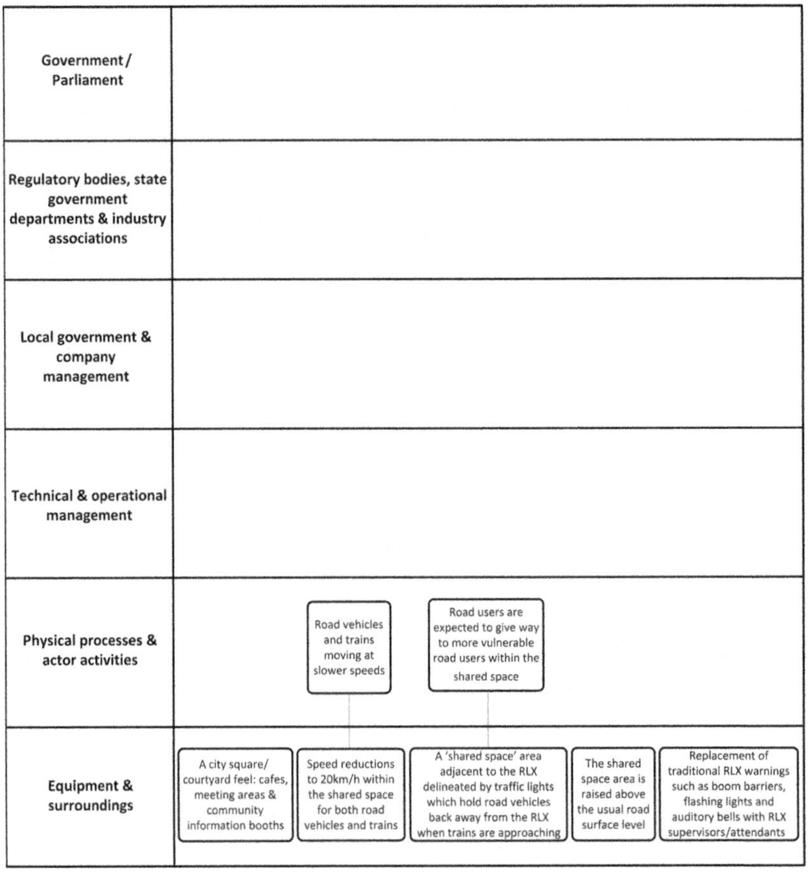

Government/Parliament					
Regulatory bodies, state government departments & industry associations					
Local government & company management					
Technical & operational management					
Physical processes & actor activities		Road vehicles and trains moving at slower speeds	Road users are expected to give way to more vulnerable road users within the shared space		
Equipment & surroundings	A city square/courtyard feel: cafes, meeting areas & community information booths	Speed reductions to 20km/h within the shared space for both road vehicles and trains	A 'shared space' area adjacent to the RLX delineated by traffic lights which hold road vehicles back away from the RLX when trains are approaching	The shared space area is raised above the usual road surface level	Replacement of traditional RLX warnings such as boom barriers, flashing lights and auditory bells with RLX supervisors/attendants

FIGURE 8.6   The Community Courtyard Crossing design concept displayed on Rasmussen's (1997) framework. RLX, rail level crossing.

(e.g. flashing lights, boom barriers and bells) are removed and replaced with human attendants. The design is intended to be implemented only in certain locations with high pedestrian traffic where the crossing is also adjacent to a train station.

The key components of the design concept are shown in Figures 8.5 and 8.6.

8.6   FINAL DESIGN CONCEPTS FOR RURAL ENVIRONMENTS

After the final design workshop, there were three concepts designed for use in rural environments: two design concepts generated through participatory workshops with stakeholders (Simple But Strong, GPS Average Speed) and one design concept generated in the final workshop with the research team (EID Crossing).

8.6.1   SIMPLE BUT STRONG

The final Simple But Strong concept represents the merging of two concepts proposed in the participatory workshops (the Speed, Expectancy, Gap concept and the initial Simple But Strong concept). The design philosophy is to use simple and low-cost features to draw attention to the upcoming rail level crossing and the danger posed, and to provide warning of an approaching train. In addition, road user speed is managed on approach to the crossing, to provide more time for drivers to recover from errors (such as seeing the train late on approach). The physical features of the Simple But Strong concept are shown in Figure 8.7, with the components described in Figure 8.8.

8.6.2   EID CROSSING

The EID concept was developed by applying the principles of EID to the design of the physical rail level crossing environment. The design features were intended to make the rail level crossing's constraints explicit to road users by emphasising the danger zone, representing the field of safe travel in the rail level crossing environment, drawing attention to the train as the key hazard and assisting both road users

FIGURE 8.7   The Simple But Strong design concept.

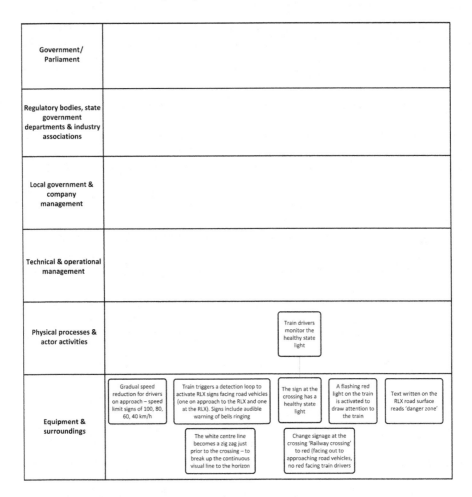

Government/ Parliament	
Regulatory bodies, state government departments & industry associations	
Local government & company management	
Technical & operational management	
Physical processes & actor activities	Train drivers monitor the healthy state light
Equipment & surroundings	Gradual speed reduction for drivers on approach – speed limit signs of 100, 80, 60, 40 km/h • Train triggers a detection loop to activate RLX signs facing road vehicles (one on approach to the RLX and one at the RLX). Signs include audible warning of bells ringing • The sign at the crossing has a healthy state light • A flashing red light on the train is activated to draw attention to the train • Text written on the RLX road surface reads 'danger zone' • The white centre line becomes a zig zag just prior to the crossing – to break up the continuous visual line to the horizon • Change signage at the crossing 'Railway crossing' to red (facing out to approaching road vehicles, no red facing train drivers)

FIGURE 8.8 The Simple But Strong design concept displayed on Rasmussen's (1997) framework. RLX, rail level crossing.

and train drivers to judge speed and distance. It achieves this by using large mirrors at the crossing to reflect an image of the train's approach, as well as a coloured zone on the approach road with a painted 'tongue' intended to represent a static field of safe travel (Gibson and Crooks 1938). Roadside markers are also placed at decreasing intervals to assist in making judgements of speed and to increase drivers' self-perceived travel speed. Finally, the livery or paint of the train itself is altered to represent a character associated with speed and strength, which increases the train's visual saliency. Finally, the design slows the train speed to 20 km/h which, combined with the traffic calming measures for road users, is intended to enable the system to better recover from errors.

The concept is shown in Figure 8.9, with its components displayed in Figure 8.10.

FIGURE 8.9   The EID concept.

FIGURE 8.10   The EID concept displayed on Rasmussen's (1997) framework. RLX, rail level crossing.

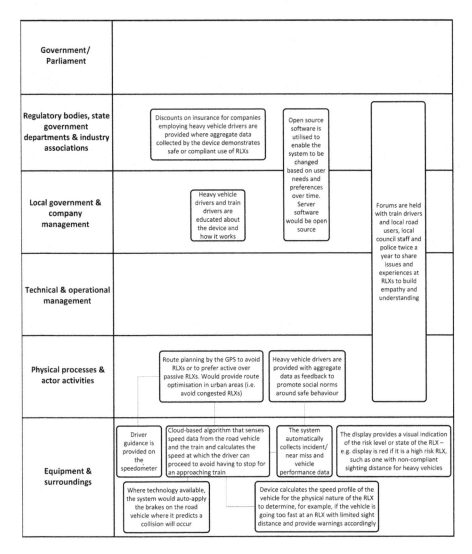

FIGURE 8.11 The GPS Average Speed design concept displayed on Rasmussen's (1997) framework. RLX, rail level crossing.

8.6.3 GPS Average Speed

The GPS Average Speed interface concept is underpinned by time-based separation and promotes efficiency and traffic flow as a means to also improve safety. It uses new and emerging technologies to provide road users (specifically drivers) with speed guidance that avoids the need to stop for an approaching train.

The components of the GPS Average Speed concept are shown in Figure 8.11. A key feature of the design is the in-vehicle assistive device, which provides speed guidance overlaid on the speedometer (see Figure 8.1). Within the interface:

- Red and green indications are used to indicate dynamic 'safe' and 'unsafe' speed zones based on the speed limit of the road and the car and train's position. The green indication is presented when the road user's speed is such that they will not encounter the train at the rail level crossing. The red indication is presented when the road user's speed is such that they will encounter the train at the rail level crossing.
- The driver is informed that the unsafe speed is imposed because of the train by a train icon appearing on the display.
- If the car's speed moves into the unsafe zone, the display flashes and an auditory tone sounds to draw attention to the display.

8.7   SUMMARY

This chapter provided an overview of the process of using the desktop evaluation findings (see Chapter 7) to refine the design concepts generated in the participatory design workshop (see Chapter 6), and to develop additional, more revolutionary design concepts for further evaluation. Importantly, it demonstrated the iterative nature of design generation and design refinement. Further, some practical lessons learned were uncovered regarding the process undertaken.

It was following the discussions at the design refinement workshop that the research team decided to develop additional expert design concepts for further evaluation. With hindsight, it may have been more appropriate for the research team to have stronger involvement in the participatory process that generated the initial designs. This would have better supported multidisciplinary learning and collaboration, and could have potentially improved the balance between academic goals and industry goals in the first instance. Although this approach was originally considered, it was not undertaken because there was some concern that expert involvement would unduly influence or bias the process. Instead, the research team adopted the role of objective facilitators. A more collaborative approach could have better balanced the need for stakeholders to have a sense of ownership over the outcomes of the process, while also exploiting the researcher teams' knowledge and expertise. Such approaches could be explored in future design efforts.

The six design concepts arising from the design process were varied in their philosophies, approaches and the extent to which they incorporated interventions across the entire rail level crossing system. To provide recommendations about which design(s) would be most effective, a series of formal evaluation processes were undertaken in the following phase of the research programme. The design concepts were tested using driving simulation (see Chapter 9) and through surveying road users to gain their perceptions of the new designs (see Chapter 10). In addition to the evaluation approaches selected for the current research programme, other forms of evaluation could also be adopted such as modelling approaches, mock-ups or focus groups.

Section IV

Evaluation of Design Concepts

9 Simulation-Based Evaluation of Design Concepts

With Contributions from:

Eryn Grant, Amanda Clacy, Miles Thomas, Nicholas Stevens, Michelle Van Mulken, Kristie Young and Christine Mulvihill

9.1 INTRODUCTION

Phase 4 of the research programme involved formal evaluation of the novel design concepts that were developed using the Cognitive Work Analysis Design Toolkit processes described in Chapters 6 through 8. This chapter will describe how designs were evaluated under experimental conditions using a driving simulator.

Simulator studies provide an ideal platform for initial evaluation of new design concepts, as it is possible to have users experience the new system design without the financial or legal obstacles that would arise when building prototypes for field testing. Indeed, preliminary concept evaluations using a simulator can provide valuable evidence to justify the subsequent cost and effort of field trials. Further, simulator studies provide a unique opportunity to test designs under genuinely experimental conditions: it is possible to use within-subjects comparisons (i.e. where all participants experience each design) and experimentally control the driving environment such that the only difference between rail level crossings is the type of control that minimises the potential for confounding factors that are likely to occur in real-world conditions (e.g. where different designs are installed in markedly different environments).

Several previous studies have used simulators to compare drivers' behaviour across different existing (Cale et al. 2013, Liu et al. 2016, Lenné et al. 2011, Rudin-Brown et al. 2012, Tey et al. 2011) and novel (Conti et al. 1998, Larue et al. 2016, Mitsopoulos et al. 2002, Tey et al. 2013) rail level crossing environments. Studies examining existing rail level crossing designs have yielded results that are highly consistent with both field observations and crash data. For instance, all types of research methods suggest that drivers exhibit lower compliance and less safe behaviour at passive rail level crossings compared with actively controlled crossings, especially those with boom barriers. However, evaluations of novel designs have focussed on in-vehicle assistive technology (e.g. in-vehicle devices that warn of approaching trains), with few published investigations of novel infrastructure or signage.

9.2   GENERAL EXPERIMENTAL METHOD

We conducted three driving simulator studies to evaluate how the proposed rail level crossing designs influenced drivers' behaviour: one focussing on the urban crossing designs and two focussing on the rural crossing designs. A fundamental aspect of the evaluations was that each new design was evaluated as a complete system, most of which included multiple new attributes, rather than evaluating each possible component change in isolation. This was done deliberately so that the evaluation reflected systems thinking principles and captured the interaction of all new elements of the system, rather than how they function individually. It should be noted, however, that only elements of designs associated with the physical rail level crossing environment were able to be tested. Some aspects of proposed designs, such as education campaigns, are not amenable to evaluation using driving simulation.

The first two studies were designed as the basic concept evaluations of how drivers responded to the new rail level crossing environments, with one study examining behaviour in simulated urban environments and the other in simulated rural environments. In these initial concept evaluations, drivers encountered the new designs under normal conditions with the infrastructure and equipment functioning as expected. The final study was a scenario-based evaluation of the rural design concepts, which extended the initial study by examining how the system functioned in the event of technology failure or when the driver was distracted on approach. The testing of failure states was important in the rural environment due to concerns regarding reliability and failure states for non-traditional active warning systems (Wullems 2011).

The same general method was used for all three evaluation studies. Experienced drivers were recruited for a single experimental session during which they completed a series of simulated drives, with each drive involving a repeated exposure to a single rail level crossing design. As in the on-road studies conducted in Phase 1 (see Chapter 4), we used a combination of objective and subjective methods to understand driving performance, including vehicle parameters, verbal protocol analysis and questionnaires measuring subjective workload and perceived usability of each design. Participants were informed that the studies related to rail level crossings, but were given no information about the new designs or the philosophies underlying them, to enable testing of intuitive responses to the novel designs.

9.2.1   DRIVING SIMULATOR

The simulator was a medium-fidelity fixed-base driving simulator, which consisted of an adjustable driver's seat, automatic transmission, Logitech G27 vehicle controls (steering wheel, accelerator and brake pedals) and three 40-inch liquid crystal display (LCD) monitors representing a 135° field of view. The LCD monitors depicted the forward and lateral road views, and operated with 1080p resolution with a 60-Hz refresh rate. The vehicle instrument panel, including speedometer and odometer, was displayed on a 9.7-inch tablet screen positioned directly behind the steering wheel. Oktal SCANeR™ studio software version 1.5 was used to programme the virtual driving environment and collect driving performance measures, including speed, acceleration and braking. For the urban design concept evaluation, an additional 7-inch tablet screen was used to simulate the in-vehicle display for the Intelligent Level Crossing design concept.

9.2.2 MEASURES

The rail level crossing designs were evaluated to assess their effects on key driving performance measures, including mean speed, speed variability, minimum and maximum speeds and stopping behaviour. Alongside these objective measures of performance, drivers' subjective workload and perceptions of usability were measured using the NASA Task Load Index (NASA-TLX; Hart and Staveland 1988) and the System Usability Scale (SUS; Brooke 1996), respectively (see Chapter 2). Participants also provided 'think-aloud' verbal protocols as a measure of situation awareness. These measures were selected to comprehensively assess drivers' behaviour and cognition, in a similar manner to the original data collection activities undertaken during Phase 1 of the research programme (see Chapter 4), and captured key processes relating to the safe negotiation of rail level crossings.

9.3 STUDY 1: URBAN DESIGN CONCEPT EVALUATION

The first study compared four urban rail level crossing designs: three novel and one existing standard design. The existing design conformed to contemporary Australian design standards and featured boom barriers, flashing lights and bells (which is the most typical infrastructure configuration for urban rail level crossings in Australia). The three novel design concepts were as follows:

1. *Comprehensive Risk Control Crossing*: This design incorporated several features in addition to the existing standard crossing, including traffic lights, advanced warning signs, in-road studs that light up when warnings are activated, additional signage to discourage queuing on the crossing, rail level crossing attendants to supervise and assist in emergencies, channelised pedestrian fencing and default closed pedestrian gates, and convenience amenities for waiting pedestrians, including a shelter and ticket machines.
2. *Intelligent Level Crossing*: This design featured identical infrastructure to the existing standard, but added an in-vehicle assistive system that provides visual and auditory cues to drivers based on the field of safe travel theory (Gibson and Crooks 1938). A green 'tongue' appears along the road representing the safe area of travel ahead. When a train is approaching the rail level crossing, a train icon appears on the display (together with an auditory alert), and a red bar appears showing the limit of the safe travel field.
3. *Community Courtyard Crossing*: This design situated the rail level crossing in a 'shared space' with a community courtyard feel, with motorists expected to give way to vulnerable road users. Traditional warnings (boom barriers, flashing lights, bells) were removed and replaced with rail level crossing attendants who supervise and assist in emergencies. Both vehicle and train speeds are reduced to 20 km/h through the shared space.

Full details of the design concepts are provided in Chapter 8. Screen shots from the urban driving simulator drives are shown in Figure 9.1.

FIGURE 9.1 Screen shots depicting the novel urban rail level crossing designs: the Comprehensive Risk Control crossing (top panel); the Intelligent Level Crossing (middle panel); the Community Courtyard crossing (bottom panel).

9.3.1 Participants

Thirty fully licenced drivers participated in the urban simulator study; however, data from one participant were discarded due to missing data. The final sample included 29 drivers (18 male, 11 female) with an average of 18.6 years' driving experience.

9.3.2 Study Design

Participants completed four simulated drives, with each drive featuring a different rail level crossing design. Within each drive, the participant drove continuously through an urban setting and encountered four rail level crossings, with trains present on two of these encounters. The drives were identical with respect to the driving environment and road layout other than the rail level crossing designs, and had equivalent traffic density and composition. The road speed limit was 60 km/h throughout, unless reduced by design. All participants experienced the existing standard rail level crossing design first, with the presentation order for the three novel designs counterbalanced between participants.

After each drive, participants were asked to complete the NASA-TLX twice: once describing their experiences when a train was present at the rail level crossing, and once describing their experiences when no train was present. They also completed the SUS and provided general comments regarding what they did and did not like about each rail level crossing design. After completing all drives, participants were asked to rank the designs in order of preference.

9.3.3 Key Findings: Comprehensive Risk Control Crossing

Analysis of vehicle performance measures revealed that speed profiles when a train was present were broadly similar on approach to the Comprehensive Risk Control Crossing compared with the existing standard (see Figure 9.2a). However, when no train was present, drivers reduced their speed noticeably in the last 20 m before the existing standard crossing, but exhibited minimal speed reduction at the Comprehensive Risk Control Crossing.

In terms of overall preference rankings, the Comprehensive Risk Control Crossing was ranked equal first with both the Intelligent Level Crossing and the existing standard crossing (see Table 9.1). Approximately one-third of participants ranked Comprehensive Risk Control as their first preference. However, in relation to usability, participants rated this crossing design ($M = 67.1$, $SD = 10.3$) as significantly lower than the existing standard ($M = 78.7$, $SD = 16.2$). Subjective workload ratings were not significantly different when comparing Comprehensive Risk Control crossing with the existing standard.

Participants' open-ended comments revealed that they particularly liked the presence of traffic lights. Consistent with this, in the verbal protocol analyses, the concept "red" emerged prominently as far as 250 m from the Comprehensive Risk Control crossing, but not any of the other designs. This suggests that the red traffic light served as an advance warning of the need to stop. Participants also appreciated the additional road markings and warning signs, as well as the designated cyclist area. However, they suggested that the attendants were unnecessary, and some felt that there were too many alerts, which made the crossing busy and distracting.

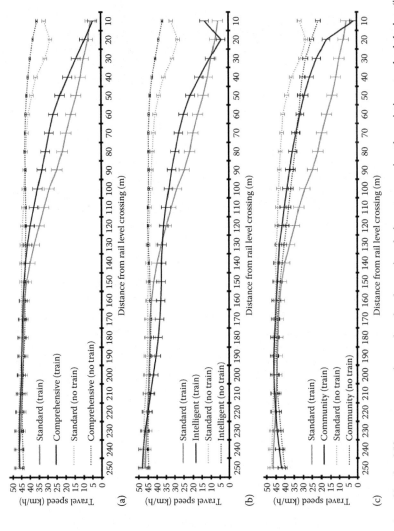

FIGURE 9.2 Speed profiles for 250 m on approach to urban level crossing designs, comparing the existing standard design (i.e. boom barriers and flashing lights) with (a) Comprehensive Risk Control crossing, (b) Intelligent Level Crossing and (c) Community Courtyard crossing. For each graph, black lines represent the novel design and grey lines represent the existing standard. Solid lines represent the average speeds when a train was approaching; dotted lines represent average speeds when no train was approaching. Error bars represent the standard error.

TABLE 9.1

Overall Preference Rankings for Urban Rail Level Crossing Designs (Simulator Study 1)

Design	First	Second	Third	Fourth	Summed Preference Score
	Preference Rank				
Intelligent Level Crossing	38%	21%	34%	7%	84
Comprehensive Risk Control	34%	24%	38%	3%	84
Existing standard crossing	24%	48%	21%	7%	84
Community Courtyard	3%	7%	7%	83%	38

Note: Preference scores were calculated by summing the rank scores, which were the inverse of the rankings (i.e. first ranking = 4, second ranking = 3, third ranking = 2 and fourth ranking = 1). As Study 1 had 29 participants, the maximum possible preference score was 116 and the lowest possible score was 29.

9.3.4 KEY FINDINGS: INTELLIGENT LEVEL CROSSING

Analysis of vehicle performance measures indicated that, when a train was present, participants exhibited more gradual speed reductions and stopped earlier at the Intelligent Level Crossing compared with the existing standard (see Figure 9.2b). When no train was present, however, participants showed minimal slowing and adopted higher approach speeds than at the existing standard crossing.

The Intelligent Level Crossing was ranked equal first in terms of preference and received the highest number of first preference rankings (see Table 9.1). It also achieved the highest usability score ($M = 79.6$, $SD = 18.9$), but this was not rated significantly higher than the existing standard ($M = 78.7$, $SD = 16.2$). Subjective workload ratings were not significantly different when comparing the Intelligent Level Crossing and existing standard.

Verbal protocol analyses revealed that the Intelligent Level Crossing alerted participants to the presence of a train much earlier than other designs; the concept 'train' reached prominence when participants were up to 250 m from the crossing, whereas for all other designs, the train concept did not emerge until within 50 m of the tracks. Participants generally described the Intelligent Level Crossing in favourable terms. They found it easy to use and liked the clear visual and auditory warnings, including the train icon on the in-vehicle display. Perceived negatives of the system included its potential for driver distraction and concerns about the reliability of the technology.

9.3.5 KEY FINDINGS: COMMUNITY COURTYARD CROSSING

Analysis of vehicle performance measures indicated that, compared with the existing standard, participants adopted slower travel speeds on approach to the Community Courtyard crossing when no train was present (see Figure 9.2c). This presumably reflects that this crossing incorporated a regulatory speed limit reduction to 40 km/h, whereas the speed limit through other crossings remained unchanged. Further, participants encountered pedestrians using the shared space, requiring them to slow or stop and give way. When a train was present, participants exhibited longer stopping times at the Community Courtyard crossing. Together these findings suggest that the Community Courtyard crossing would create greater travel delays for drivers compared with the existing standard.

The Community Courtyard crossing was rated as the least preferred crossing (see Table 9.1) and as having the lowest usability ($M = 42.3$, $SD = 25.0$), which was significantly lower than all other urban designs. Several aspects of subjective workload were also elevated for this design, compared with the existing standard. Specifically, participants reported significantly higher frustration levels when negotiating the Community Courtyard crossing, both with and without a train present. Participants also reported significantly higher temporal demands when a train was present and significantly higher mental demands when no train was present.

Participants described the Community Courtyard crossing as visually appealing and appreciated the certainty provided by traffic lights, but expressed mixed feelings regarding the shared space. Participants noted that the shared space environment effectively 'forces' drivers to adopt lower travel speeds, but some found it confusing and felt unsafe with the proximity and lack of delineation between different crossing users (i.e. pedestrians, road vehicles and trains). Although vehicle data showed participants slowing on approach, subjective data revealed that many did not realise they were approaching a rail level crossing; rather, the verbal protocol analyses suggested that drivers focussed predominantly on the traffic lights and were unaware of the level crossing until they saw the train.

9.3.6 Summary of the Urban Design Evaluation Findings

Of the three novel urban rail level crossing designs evaluated, the Intelligent Level Crossing received the highest number of first preference rankings (although overall preference scores were equal with the Comprehensive Risk Control and existing standard crossings). The Community Courtyard crossing was least preferred. The Community Courtyard crossing also exhibited potential for traffic delays, with longer stopping times when a train was present and slower travel speeds when no train was present. In contrast, both the Comprehensive Risk Control and Intelligent Level Crossing designs showed noticeably less speed reduction when no train was present, compared with the existing standard design. The most likely explanation for this is that both new designs effectively provided confirmation that it was safe to proceed (i.e. green traffic light; field of safe travel representation), whereas in the existing standard design, participants felt the need to slow on approach to ensure that the crossing was inactive and that no trains were approaching. Consistent with this, drivers also reported appreciating the inclusion of traffic lights in the Comprehensive Risk Control and Community Courtyard crossings, as these provide greater decision-making certainty.

Across all designs, participants expressed preference for delineation between different road user groups (e.g. the separate cyclist lane in the Comprehensive Risk Control crossing) and were uncomfortable with removing barriers to create a 'shared space' environment. Human level crossing attendants were perceived as auxiliary and unnecessary; however, in the proposed designs, these attendants were primarily intended to assist in emergency situations, which were not tested in this initial concept evaluation.

9.4 STUDY 2: RURAL DESIGN CONCEPT EVALUATION

The second simulator study compared five rail level crossing designs: three novel and two existing standard designs. Both the existing designs conformed to contemporary Australian design standards. One was an active crossing, featuring flashing lights and bells but no boom barriers, and the other was a passive crossing, featuring a Give Way (yield) sign. These two standard designs were adopted in this study as they represent the two main types of existing crossings that would likely be prioritised for upgrade. Further, they enable comparisons to be made between existing and proposed passive designs, as well as between existing and proposed active designs.

The three novel design concepts tested in the rural evaluation were as follows:

1. *Simple But Strong*: This design features a lower cost active warning sign at the crossing, as well as an active advance warning sign on approach. Both the signs incorporate flashing lights and audible bells. There is a gradual reduction of the speed limit on approach, from 80 to 60 km/h and finally 40 km/h. Coloured road markings at the crossing visually delineate the 'danger zone'.
2. *Ecological Interface Design*: This design features large mirrors at the crossing, which reflect the image of an approaching train to signal its presence and relative distance. It also incorporates several elements designed to slow drivers on approach, including a coloured zone on approach with a 'tongue' intended to represent a static field of safe travel (Gibson and Crooks 1938), as well as roadside markers placed at decreasing intervals (to increase the driver's self-perceived travel speed). Finally, the train itself is altered to be more salient and visually imposing, representing a character associated with speed and strength, and at the same time, its travel speed through the crossing is reduced to 20 km/h.
3. *GPS Average Speed*: This design features identical infrastructure to the existing standard, but includes an in-vehicle assistive system that alerts the driver to the presence of an approaching train and provides guidance on safe travel speeds. 'Safe' travel speeds are those that do not exceed the posted speed limit and would avoid conflicts with a train at a level crossing. If the user is travelling at a speed that would require them to stop for a train to avoid a collision, the system provides a brief auditory alert and flashes red on the speedometer.

Again, full details of these design concepts are provided in Chapter 8. Screen shots from the rural driving simulator drives are shown in Figure 9.3.

FIGURE 9.3 Screen shots depicting the three novel rural rail level crossing designs evaluated in Simulator Study 2 and Study 3: the Simple But Strong crossing (top panel); the Ecological Interface Design crossing (middle panel); the GPS Average Speed interface (bottom panel).

9.4.1 PARTICIPANTS

Thirty fully licenced drivers (21 male, 9 female) aged between 20 and 55 years participated in the rural simulator study. Participants had an average of 14.8 years' driving experience.

9.4.2 STUDY DESIGN

Participants completed five simulated drives, with each drive featuring a different rail level crossing design but otherwise matched in terms of road environment and traffic conditions. The road speed limit was 100 km/h with an 80 km/h reduction on approach to all rail level crossings, sometimes reduced further depending on the design presented. Presentation order for the different drives was counterbalanced between participants. Within each drive, the participant drove continuously through a rural setting and encountered five rail level crossings: two with a train present and three with no train present. The first rail level crossing was always in an inactive state (i.e. train absent), with the order of subsequent train-present versus train-absent exposures counterbalanced between drives. As in the first study, participants completed the NASA-TLX and SUS after each drive, and were asked to describe what they liked and disliked about each design. At the end of the study, participants ranked the designs in order of preference.

9.4.3 KEY FINDINGS: SIMPLE BUT STRONG

Analysis of travel speeds indicated that overall travel speeds were slowest for the Simple But Strong crossing, regardless of whether a train was present (see Figures 9.4a and 9.5a). This was associated with significantly longer travel times to negotiate the crossing, compared with standard active and passive crossings, for both train-present and train-absent conditions. Due to the greater speed reductions, drivers also exhibited greater variability in their travel speeds (i.e. higher standard deviation of speed during the approach segments). Although participants generally complied with the posted speed reductions, many explicitly stated that they would not do so under real-world conditions.

In terms of overall preference rankings, Simple But Strong was the least preferred design, with no participants rating it as their top preference (see Table 9.2). Overall usability ratings for the Simple But Strong crossing ($M = 62.3$, $SD = 25.3$) were also significantly lower than those of both standard active ($M = 78.8$, $SD = 25.3$) and standard passive ($M = 80.0$, $SD = 18.4$) crossings. Participants especially rated the Simple But Strong design as being unnecessarily complex and less easy to use. The only aspect of workload that differed with this design was physical workload, which was rated as significantly higher than the standard active (but not the standard passive) crossing when a train was present. However, the practical significance of this effect was small, as all designs were rated as having relatively low workload.

Subjective data revealed that participants most liked the active warning signs, especially the advance active warning signs. Responses to the coloured 'danger zone' were mixed, with some participants liking this feature and others expressing concern that it could divert drivers' attention away from scanning for trains. Drivers generally disliked the speed reduction and number of speed signs, although a minority appreciated the guidance as it allowed them extra time to prepare for potential hazards.

9.4.4 KEY FINDINGS: ECOLOGICAL INTERFACE DESIGN CROSSING

Analysis of travel speeds on approach to the Ecological Interface Design crossing revealed different patterns for the train-present and train-absent conditions. When a train was present, drivers showed large speed reductions at the Ecological Interface Design crossing, with approach speeds being slower than those at the standard active crossing (see Figure 9.4b) and slightly slower than those at the standard passive crossing (see Figure 9.5b). Conversely, when no train was present, drivers exhibited minimal speed reductions on approach to the Ecological Interface Design crossing. Approach speeds for the Ecological Interface Design were comparable to those for the standard passive crossing and significantly higher compared with the standard active crossing, as drivers showed significant speed reductions on approach to the active crossing even when no train was approaching.

In terms of overall preference rankings, the Ecological Interface Design crossing was the most preferred of the novel designs and was the second ranked overall, just behind the standard passive crossing (see Table 9.2). Interestingly, however, usability ratings for the Ecological Interface Design crossing ($M = 62.9$, $SD = 25.9$) were

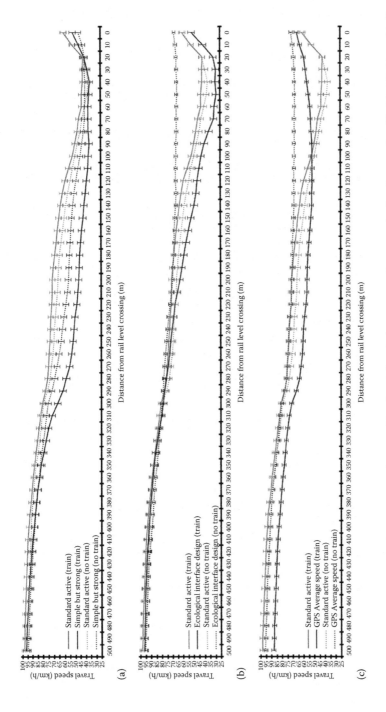

FIGURE 9.4 Speed profiles for 500 m on approach to rural level crossing designs, comparing a standard active crossing (flashing lights and bells, no boom barriers) with (a) Simple But Strong, (b) Ecological Interface Design and (c) GPS Average Speed. For each graph, black lines represent the novel design and grey lines represent the existing standard. Solid lines represent the average speeds when a train was approaching; dotted lines represent average speeds when no train was approaching. Error bars represent the standard error.

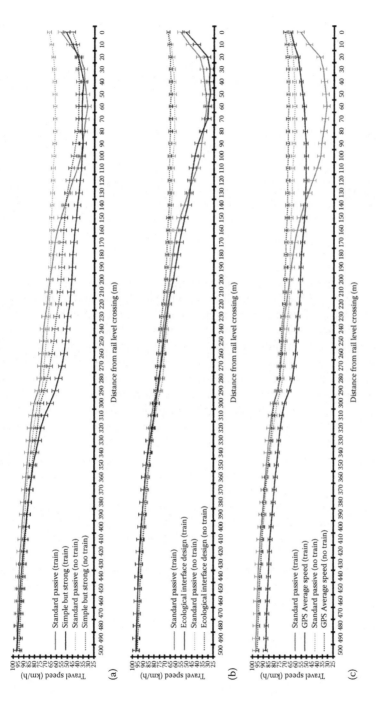

FIGURE 9.5 Speed profiles for 500 m on approach to rural level crossing designs, comparing a standard passive crossing (i.e. Give Way sign only) with (a) Simple But Strong, (b) Ecological Interface Design and (c) GPS Average Speed. For each graph, black lines represent the novel design and grey lines represent the existing standard. Solid lines represent the average speeds when a train was approaching; dotted lines represent average speeds when no train was approaching. Error bars represent the standard error.

TABLE 9.2

Overall Preference Rankings for Rural Rail Level Crossing Designs (Simulator Study 2)

Design	Preference Ranking					Summed Preference Score
	First	Second	Third	Fourth	Fifth	
Standard passive	40%	13%	17%	–	30%	100
Ecological Interface Design	27%	20%	17%	20%	17%	96
GPS Average Speed	7%	40%	7%	37%	10%	89
Standard active	27%	3%	30%	10%	30%	86
Simple But Strong	–	23%	30%	33%	13%	79

Note: Preference scores were calculated by summing the rank scores, which were the inverse of the rankings (i.e. first ranking = 5, second ranking = 4, third ranking = 3, fourth ranking = 2 and fifth ranking = 1). As Study 2 had 30 participants, the maximum possible preference score was 150 and the lowest possible score was 30.

significantly lower than those for both the standard active ($M = 78.8$, $SD = 25.3$) and standard passive ($M = 80.0$, $SD = 18.4$) designs, with participants finding the design more complex, requiring more technical support and less easy to use than existing standards. Several aspects of workload were also elevated. Compared with the standard active crossing, participants rated the Ecological Interface Design crossing as more frustrating (regardless of whether a train was approaching), as having higher physical workload when a train was approaching and as having higher temporal demands when no train was present. Participants also rated the Ecological Interface Design crossing as requiring higher effort than the standard passive crossing when no train was present. As with the Simple But Strong results, even when comparisons were statistically significant, the practical significance is small as most ratings were at the extreme lower end of the scale.

Participants' verbal responses indicated that they liked the train colour and salience. Other aspects of the design received mixed responses; for instance, some drivers liked the road markings and roadside reflector poles as these attributes encouraged slower approach speeds, whereas other drivers found the infrastructure overwhelming and were unsure how to respond. The mirrors also generated confusion for some participants, which reflects a limitation of the simulation graphics: the driver-facing mirror was rendered as a flat grey sign, so several participants were confused as to its purpose and did not use it for information seeking. The participants who did realise the mirror's function reported finding it useful.

9.4.5 KEY FINDINGS: GPS AVERAGE SPEED

When no train was approaching the level crossing, drivers showed minimal speed reductions when using the GPS Average Speed interface. Both travel speeds and overall efficiency (i.e. time taken to negotiate the crossing) were similar to the standard passive crossing (see Figure 9.5c), and were faster than the standard active crossing (see Figure 9.4c). The GPS interface remained highly efficient when a train was present, with the time required to negotiate the crossing being similar to the standard active crossing and faster than all other designs. Although the overall efficiency was similar for the GPS Average Speed and standard active crossings, the speed profiles were different: when using the GPS interface drivers slowed earlier and to a lesser extent, maintaining a more constant pace with less variability in their approach speed over the last 250 m before the crossing.

Analysis of the time spent with the GPS Average Speed system in the 'warning state' revealed substantial individual variability. The earliest point at which the system could be activated within the simulated scenario was 2 km from the crossing. The average duration spent in the red 'unsafe speed' zone was 30% of this 2 km approach to the rail level crossing, with individual participants ranging from no activations to having it active for 90% of the time. Notably, most participants had the system alternate between 'unsafe' and 'safe' states several times, with an average of two to three discrete warning activations. This suggests that some drivers may have been attempting to stay close to the upper end of the 'safe' speed zone. Similar behaviour has been observed with intelligent speed adaptation systems (Regan et al. 2005).

In terms of overall preference rankings, the GPS Average Speed interface was ranked third overall and second of the novel designs, behind the standard passive and Ecological Interface Design crossings and just ahead of the standard active crossing (see Table 9.2). Usability ratings for the GPS Average Speed interface ($M = 73.5$, $SD = 22.1$) were not significantly lower than those of either the standard active or passive crossings. Overall workload measures were also not different to existing standards, except for physical workload, which was rated as being slightly higher for GPS Average Speed compared with the standard active (but not the standard passive) crossing when a train was present. This presumably reflects the fact that drivers had an additional display to monitor when a train was approaching.

Verbal responses revealed both positive and negative assessments of the GPS Average Speed interface. Participants liked that they could use it to avoid stopping for a train; however, not all participants intuitively understood the information that was being conveyed. We deliberately provided only a brief description of the system function to participants before the drive, to mimic a situation in which a driver hires or buys a new car and is told that it has a specific assistive system but is not given details on how the system functions. Some participants also expressed negative attitudes towards the interface, indicating that the speed reductions required were too great and that they found the system distracting when the red alert was flashing.

9.4.6 Summary of the Rural Design Evaluation Findings

Among the three novel rural designs tested in the initial concept evaluation, the Ecological Interface Design was the most preferred by participants even though it was rated as less usable than existing standard crossing designs. The GPS Average Speed interface was generally rated favourably in terms of both usability and subjective preference, whereas the Simple But Strong crossing was rated lowest in both usability and preference scores. Participants seemed to especially dislike the speed reductions in the Simple But Strong design, but liked the addition of coloured road markings in both the Simple But Strong and Ecological Interface Designs crossings. The popularity of the Ecological Interface Design also highlights the potential for further work to investigate increasing the conspicuity of the train to improve its detectability.

9.5 STUDY 3: SCENARIO-BASED EVALUATION

The first two simulator studies provided initial design concept evaluations, which represent the first required step in testing new designs with intended system users. To further explore the functioning of the new design concepts, we conducted a follow-up study to evaluate how drivers responded to the novel rail level crossing treatments in safety-critical situations. We refer to this as 'scenario-based' evaluation, as we placed participants in a specific scenario designed to alter the nature of their interaction with the rail level crossing system.

To devise a list of scenarios, we reviewed the findings from the Phase 1 data collection to identify issues and concerns that road users have when interacting with level crossings. We then consulted with rail and road industry stakeholders to obtain their input on issues that contribute to collisions and near misses at rail level crossings. From these activities, we identified the following two issues:

1. *Technology reliability and failure*: Concerns about signal reliability were frequently expressed by research participants in Phase 1, especially those in rural areas. Participants in the first two simulator studies also raised concerns about the reliability of the active warnings, especially the GPS-based in-vehicle interfaces. Rail industry stakeholders noted that technology failure can include complete failure (i.e. no warning activation at all), as well as delayed or mistimed activation of warnings (i.e. the full warning time is not provided to road users).

2. *Driver distraction*: Two aspects of distraction were identified as having relevance to level crossings. First, participants in the initial concept evaluations expressed that they found some aspects of the new designs distracting, which raises issues about overload and the user's ability to process all information. Second, stakeholders expressed concern regarding the increased proliferation of in-vehicle technology, including both in-vehicle information and assistance systems and portable electronic devices (e.g. cellular phones). In-vehicle distractions have the potential to divert attention away from the level crossing environment, again impacting the driver's ability to process crucial information within the environment.

While technology failure is most problematic in rural environments, distraction can occur in any environment. For efficiency, both scenarios were tested in the rural environment. The study design was broadly similar to Study 2, except that only the standard active crossing was directly compared with the new designs, because the passive standard does not include any technology or devices that would have an equivalent 'fail' state.

9.5.1 Participants

Twenty-five fully licenced drivers (8 male, 17 female) participated in the simulator study. Participants had an average of 15.0 years' driving experience.

9.5.2 Study Design

Participants completed five simulated drives, with each drive featuring a different rail level crossing design. Four of the five drives – specifically those featuring the standard active, Simple But Strong, Ecological Interface Design and GPS Average Speed designs – were fully matched in terms of road environment and traffic conditions, and included both the 'system failure' and distraction scenarios. Each of these drives included the following four exposures to rail level crossings:

1. Train approaching, normal conditions
2. Train approaching, system failure state
3. No train approaching, normal conditions
4. No train approaching, in-vehicle distraction

The first rail level crossing exposure within each drive always had a train approaching under normal conditions, so that participants could experience the rail level crossing system in its optimal functioning state. The order of the other events was counterbalanced between conditions.

The simulated drive featuring the standard passive crossing was shorter, with only three crossings encountered, as it did not include the system failure state exposure. This was always the third drive in the sequence and was included to prevent participants from developing strong expectations about the length and nature of drives (i.e. that all would have four crossings, one of which would involve technology failure).

As in the concept evaluation studies, participants completed the NASA-TLX and SUS for each design, verbally described what they liked and disliked about each design and finally ranked the designs in order of preference.

9.5.2.1 Distraction Scenario

The impact of in-vehicle distraction was assessed using a secondary visual detection task. This task involved a small red dot appearing on the dashboard at pre-programmed pseudo-random intervals during the drive. The participant's task was to press a button on the steering wheel as quickly as possible, whenever they detected the stimulus appearing. A similar task has been validated in previous research and has been found to induce distraction-related errors in real driving (e.g. Young et al. 2013a).

The distractor occurred in three contexts: on approach to a rail level crossing, while driving on a straight stretch of road and while negotiating a corner. The current analyses focussed on response times when the participant was approaching a rail level crossing. The key question was whether response time differed between crossing designs. Systematic differences in response times between crossings would suggest that some designs impose greater attentional demands (as if attentional demands are low, participants should be able to respond quickly).

9.5.2.2 System Failure Scenario

The impact of system failure was evaluated by exposing participants to each rail level crossing in a situation that involved a train approaching, with the warnings being unavailable or functioning in a suboptimal manner. Due to the diverse characteristics of the designs tested, each system had a unique failure state:

- *Standard active crossing*: The failure state involved a simulated late activation of the warnings. The flashing lights and bells were triggered late, with a delay of 15 s, giving participants less time to respond. Note that according to Australian Standards, the minimum warning time should be 25 s.
- *Simple But Strong*: The failure state involved a simulated power failure, a response to which had been proactively designed into the initial concept. In the event of a power failure, the active portion of the sign is covered by a passive sign that indicates the sign has failed and that drivers should give way to trains.
- *Ecological Interface Design*: The failure state involved simulated vandalism, with the mirrors defaced by graffiti to obscure the view of approaching trains.
- *GPS Average Speed*: This failure state involved a simulated loss of connection between the in-vehicle device and the cloud-based software providing the information about the train's approach. This was presented to participants as the device initially activating to provide speed guidance, with an auditory tone provided, but failing after approximately half a second (600 ms) and the visual speed guidance disappearing from the speedometer display.

9.5.3 Key Findings

As shown in Table 9.3, participants' preference rankings for the three novel designs were consistent with the initial rural design concept evaluation study: Ecological Interface Design was the most preferred, just ahead of GPS Average Speed, with Simple But Strong the least preferred design. The relative rankings of the two existing standards were reversed, with the standard active crossing emerging as the most preferred overall. This is likely because participants had greater exposure to the active crossing, compared with the passive crossing, in Study 3.

Analysis of distractor response times on approach to the rail level crossings revealed that most were approximately 1–1.5 s. Response times longer than 4.5 s were considered 'lapses' and removed from the analysis. There were systematic differences between conditions, with response times being significantly longer for both the GPS Average Speed interface ($M = 1.51$ s, $SD = 0.79$) and the Ecological Interface Design crossing ($M = 1.38$ s, $SD = 0.77$) compared with the standard active crossing ($M = 1.11$ s, $SD = 0.43$). Consistent with this, participants also reported significantly higher temporal demands on the NASA-TLX for these two designs in the train-absent condition (i.e. the condition in which the distractor was presented), but not in the train-present condition. Notably, both the GPS Average Speed and the Ecological Interface Design have only passive elements at the crossing itself, which

TABLE 9.3

Overall Preference Rankings for Scenario-Based Evaluation of Rural Rail Level Crossing Designs (Simulator Study 3)

Design	Preference Ranking					Summed Preference Score
	First	Second	Third	Fourth	Fifth	
Standard active	40%	8%	28%	4%	20%	86
Ecological Interface Design	20%	28%	16%	20%	16%	79
GPS Average Speed	12%	24%	28%	24%	12%	75
Standard passive	28%	12%	12%	20%	28%	73
Simple But Strong	–	28%	16%	32%	24%	62

Note: Preference scores were calculated by summing the rank scores, which were the inverse of the rankings (i.e. first ranking = 5, second ranking = 4, third ranking = 3, fourth ranking = 2 and fifth ranking = 1). As Study 3 had 25 participants, the maximum possible preference score was 125 and the lowest possible score was 25.

places a greater onus on drivers to make their own visual checks for trains. This would be especially true in the current study, where participants may have already experienced a failure state where warnings were unavailable (Ecological Interface Design mirrors) or unreliable (GPS alert). These two design concepts are also the least similar to the existing standards, thus potentially requiring more attention from drivers, which could account for longer response times to the distractor.

Distractor response times on approach to the Simple But Strong crossing ($M = 1.21$ s, $SD = 0.41$) were not significantly different from the standard active crossing. Two design features of the Simple But Strong crossing likely contributed to participants' ability to respond faster: the design included active warning signs at and on approach to the crossing, reducing the need for drivers to make visual checks on approach; and the speed limit was substantially reduced, so drivers were travelling slower. Although temporal demands were not an issue for the Simple But Strong crossing, participants reported significantly higher frustration levels when no train was present, as they deemed the speed reduction 'unnecessary' in the absence of a train.

The introduction of the failure mode for each design did not result in participants failing to stop for approaching trains, suggesting that participants were not simply relying on warnings and were actively checking for trains. This may also be explained by a relatively wide field of view and unobstructed vision available to participants, with approaching trains visible in peripheral vision well in advance of the crossing.

Interestingly, including the system failure scenario seemed to improve participants' perceptions of the Ecological Interface Design. Although it was the most preferred novel design in Study 2, participants rated it as having lower usability and higher workload demands than the existing standards. In contrast, in the scenario-based evaluation, there was no significant difference in usability scores when comparing the Ecological Interface Design ($M = 65.9$, $SD = 6.3$) with the standard active crossing ($M = 80.0$, $SD = 18.7$). Although the mean usability score for the Ecological Interface Design crossing was low, this was skewed by a few extremely low ratings, with the median score (75) being considerably higher. The only aspect of workload that differed significantly between the Ecological Interface Design and the standard active crossing was performance, which was rated as significantly *better* (i.e. closer to 'perfect') with the new design.

Results for the GPS Average Speed interface were similar to Study 2: overall usability ratings ($M = 70.3$, $SD = 18.8$) were not significantly lower than those for the standard active crossing. Aside from temporal demands in the train-absent condition (noted previously), the only other aspect of workload that differed between the GPS Average Speed interface and the standard active crossing was physical demands, which were significantly higher when using the GPS Average Speed interface and a train was approaching. Again, this is consistent with the results of Study 2.

Finally, as with Study 2, usability of the Simple But Strong crossing ($M = 63.5$, $SD = 12.5$) was rated as significantly lower than that of the standard active crossing, and mental workload was rated as significantly higher, regardless of train presence.

9.6 SUMMARY

The results from the three driving simulator studies provided key insights into the strengths and weaknesses of the proposed novel designs compared with existing standard crossings. Adopting a holistic approach to the evaluation of the designs had benefits in relation to gaining overall measures of driver behaviour and subjective experience. However, this approach to evaluation did have some limitations in relation to providing objective evidence of the efficacy of individual components within the designs. Assessments of the components with the most promising benefits were instead based on researcher judgement, taking into account all available evidence.

Another interesting aspect to this evaluation was the ability of simulation to fully represent the intended design features (e.g. the Ecological Interface Design mirrors). Potentially, a higher fidelity 'mirror' could have been programmed into the simulator, using more sophisticated graphics, yet the time and cost to achieve this may have made the study unviable. Instead, this highlights that some design aspects may only be validly tested in field trials.

Finally, although we tested the rural design concepts under adverse scenarios, there are a multitude of scenarios that could have been tested with either the urban or the rural design. For example, scenarios could be based on the key risks identified from Phases 1 and 2 of the research (see Chapter 5), which included task under-engagement in monotonous environments, congestion (leading to drivers queuing over crossings), frustration (due to situations such as extended warning times), expectancy and time pressure. Some of these risks are difficult to replicate in a driving simulator; for example, to create a strong expectancy would require a very large number of encounters, meaning that participants are likely to become fatigued throughout the experiment. An important question therefore is the extent to which all scenarios should be tested, or whether such detailed testing should be attempted only for designs planned for implementation after further evaluation efforts.

A further evaluation study is described in Chapter 10 to address a key limitation of driving simulator studies, namely, that they can only be used to assess the performance of drivers. In this study, we solicited evaluation feedback from representatives of all road user types to better understand how different road user groups might respond to the new designs and to explore similarities and differences of opinion in relation to the designs between different road users.

10 Survey-Based Evaluation of Design Concepts

With Contributions from:
*Tony Carden, Nicholas Stevens
and Miles Thomas*

10.1 INTRODUCTION

This chapter will describe the evaluation of the novel rail level crossing designs by a variety of road users. This study was designed to build on the experimental driving simulator evaluation that was described in Chapter 9, which focussed exclusively on car drivers as participants. The major advantage of simulation is that it allows road users to experience the rail level crossing in a similar fashion to how they would experience it in the real world. For practical reasons, however, it is difficult to conduct simulations across different road user groups as each group requires physically different simulators, which are expensive to build, maintain and run. For this reason, most land transport research groups have only car simulators, with less research being conducted using heavy vehicle, motorcycle, bicycle or pedestrian simulators. It is also time and labour intensive to conduct simulator studies, and the participant sample is typically limited to those who live or work near the research centre.

An alternate approach to overcome this gap is to expose potential users to simulated scenarios via an online survey study. To achieve this, videos were created using the simulated drives from the studies described in Chapter 9, but adapted to mimic the perspectives and actions of different road users. For example, the videos depicting pedestrian experiences represented the path of someone walking through the designated pedestrian section of the rail level crossing (see Figure 10.1). Survey respondents were asked to describe what they liked and disliked about each design, and to compare the designs with respect to the key criteria such as safety, compliance, efficiency and preference. These criteria were adapted from the values and priorities identified in the Work Domain Analysis (see Figure 5.1), with preference added as an overall measure of usability.

FIGURE 10.1   Still images from the pedestrian videos, depicting the Comprehensive Risk Control crossing (top panel), Intelligent Level Crossing (middle panel) and Community Courtyard crossing (bottom panel).

10.2   SURVEY METHOD

A set of nine surveys was constructed online using SurveyMonkey® (www. surveymonkey.com), with each version tailored to one of five road user groups (car drivers, heavy vehicle drivers, motorcyclists, cyclists and pedestrians) residing in one of two locations (urban or rural Australia). This design allowed 10 possible combinations of location and road user type; however, no rural pedestrian survey was offered as the rural designs all depicted high-speed roads in relatively remote locations, so it was deemed unlikely that pedestrians would use these crossings.

Each survey contained the following three sections:

- *Demographic information*: Respondents were asked to provide background information, including age, gender, travel habits and whether they have any professional experience related to rail level crossing safety.
- *Animated simulations of level crossing designs*: Respondents viewed a series of video clips depicting simulated run-throughs of several rail level crossing design concepts from the perspective of the relevant road user. For each design, they received a short written statement drawing attention to the key aspects of the design and were asked to describe what they liked and disliked about the design.
- *Paired comparisons*: Respondents completed a series of forced-choice paired comparisons, during which two still images depicting different design concepts were displayed side by side. All possible pair combinations were included, each on a separate page, with combinations presented in random order. Each image included a caption with the design name and

a hyperlink allowing respondents to re-watch the animated run-through video. For each pair, respondents answered the following four forced-choice questions:

- *Safety*: Which crossing would have less collisions between trains and drivers/heavy vehicle drivers/motorcyclists/cyclists/pedestrians?
- *Compliance*: At which crossing would drivers/heavy vehicle drivers/motorcyclists/cyclists/pedestrians be most likely to stick to the rules?
- *Efficiency*: Which crossing would be faster to get through for drivers/heavy vehicle drivers/motorcyclists/cyclists/pedestrians?
- *Preference*: Which crossing do you think drivers/heavy vehicle drivers/motorcyclists/cyclists/pedestrians would prefer to use?

Nearly 200 respondents completed the online surveys across all road user groups, with participants self-selecting their version based on their geographic location and their most frequently used mode of transport while traversing a rail level crossing. The surveys were advertised to Australian residents online via social media and special interest groups and were open to the public for a 1-month period in late 2016.

10.3 DATA ANALYSIS

Paired comparison analysis was used to aggregate and analyse responses to the forced-choice questions. This involved creating one comparison table for each of the four comparison questions, for each road user group. An example is shown in Table 10.1.

For each pair of design concepts (e.g. Comprehensive Risk Control crossing compared with Intelligent Level Crossing), the sum of responses for the preferred design concept in each comparison question was divided by the sum of responses for the non-preferred design concept. The resulting ratio was entered into the paired comparison table, annotated with a letter to indicate which of the two design concepts was preferred. Preference ratios for each design concept were then summed to derive an overall ranking, as shown in the two rightmost columns of Table 10.1.

TABLE 10.1

Example Paired Comparison Ranking Table (Urban Car Drivers, Safety Criterion)

Design Concept	Paired Comparisons			Overall Score	
	B	C	D	Total	Rank
A. Standard active	B 13.8	C 4.9	A 1.2	1.2	3
B. Comprehensive Risk Control	–	B 4.9	B 6.4	25.1	1
C. Intelligent Level Crossing	–	–	C 1.4	6.3	2
D. Community Courtyard	–	–	–	0	4

10.4   URBAN DESIGN EVALUATIONS

10.4.1   Urban Car Drivers

Fifty-nine respondents (24 female, 35 male) aged between 18 and 74 years completed the urban car drivers' survey. Of these, 22% had professional experience relevant to rail level crossing safety, with an average of 9.2 years' work experience (range: 2–45 years).

Most respondents were aged between 35 and 54 years (66%), held a full/open driver's licence (98%) and lived in the state of Victoria (68%). The remainder of the sample resided in Queensland (15%), New South Wales (12%), Western Australia (3%) and South Australia (2%). The sample comprised frequent users of rail level crossings, with 32% reporting that they crossed level crossings several times a month, 29% several times a week, 17% every day and 3% several times a day. Respondents reported driving an average of 248 km each week ($SD = 210$) with most nominating commuting (51%) as the main purpose for most of their driving.

Paired comparison analysis indicated that car drivers rated the Comprehensive Risk Control crossing highest for safety and compliance, whereas they rated the Intelligent Level Crossing highest for efficiency and preference. The Community Courtyard crossing was ranked lowest on all the four criteria.

Car drivers' responses to the open-ended questions regarding the Comprehensive Risk Control crossing were mixed, with most respondents noting several features that they liked but also the aspects that they disliked. Only a small number reported that they did not like anything about the design. In general, respondents liked the additional warnings and controls, although no specific element of the design emerged as the most preferred. The main criticisms were that it was 'overkill' with too many elements, some of which were deemed unnecessary. Many drivers stated that the attendants at the crossing were unnecessary, and potentially in danger, but a small number liked the attendants and the fact that having human attendants creates job opportunities.

Responses regarding the Intelligent Level Crossing were mostly positive; however, a sizeable minority of respondents were strongly opposed to this design. Car drivers commented positively on the fact that the system provided additional warnings, provided auditory alerts and could be integrated with existing in-vehicle systems. Criticisms raised were that the system could be distracting, drivers may miss the alerts, the system may not be compatible with all vehicles and many drivers do not use in-vehicle navigation systems or do not use them regularly. Several expert respondents suggested that the information about the direction of train approach was unnecessary, and that alerts could become extremely complex (e.g. if the crossing had multiple tracks, or if the boom gates malfunctioned).

Responses to the Community Courtyard crossing were mixed, with respondents commenting positively on the aesthetics, open design, the inclusion of traffic lights, lowered speed limits, delineation provided by the raised road section and the fact that traffic was stopped far back from the rail level crossing. However, many respondents expressed concerns about safety and the potential for confusion, with several comments about the lack of boom barriers and explicit signage indicating that a rail level crossing was ahead.

10.4.2 Urban Heavy Vehicle Drivers

Four male respondents completed the urban heavy vehicle drivers' survey, comprising three current professional drivers aged between 25 and 54 years, all of whom lived in Victoria, and one retired driver from Queensland. Most reported driving approximately 1,000 km each week and encountering rail level crossings several times per day.

Paired comparison analysis found that urban heavy drivers rated the Comprehensive Risk Control crossing highest on all the four criteria (safety, compliance, efficiency and preference). The standard active crossing was ranked second for most criteria, with the Community Courtyard and Intelligent Level Crossing ranked equal third.

Consistent with the paired comparisons, open-ended comments were mostly positive regarding the Comprehensive Risk Control crossing. Heavy vehicle drivers reported liking the additional safety features, clear warnings and separation of vehicles, cyclists and pedestrians. One driver was concerned that the design was labour intensive (i.e. because of the attendants and extra controls that need to be installed and maintained), and another suggested that advanced warnings could be added to improve the design.

Most heavy vehicle drivers also commented positively on the Intelligent Level Crossing for providing additional warnings. However, one driver expressed his concern about the cost of such systems and another regarding system malfunctions.

Heavy vehicle drivers' responses to the Community Courtyard crossing were mixed. The positive aspects noted were that traffic was stopped well back from the level crossing, attendants could assist in situations where signals malfunctioned and having attendants increased employment. However, drivers also expressed concerns regarding the lack of boom gates and warning signs, the potential for delays and the cost and labour required.

10.4.3 Urban Motorcyclists

Twenty-two respondents (1 female, 21 male) aged between 18 and 64 years completed the urban motorcyclists' survey, of whom three had professional experience relevant to rail level crossing safety. Nearly half of the respondents were aged between 35 and 44 years (45%) and all but one held a full/open motorcycle licence (95%). Respondents lived in Queensland (41%), Victoria (23%), New South Wales (18%), Western Australia (14%) and Tasmania (5%).

The sample comprised relatively frequent users of rail level crossings, with 27% reporting that they crossed level crossings several times a month, 23% several times a week, 14% every day and 5% several times a day. Respondents reported riding an average of 194 km each week ($SD = 137$), with most nominating commuting (59%) as the main purpose for most of their riding.

Paired comparison analysis indicated that motorcyclists rated the Comprehensive Risk Control crossing highest for safety and compliance and equal first for preference (alongside the standard active crossing). The standard active crossing was rated highest for efficiency, whereas the Community Courtyard crossing was ranked last or equal last on all the criteria. Motorcyclists were not asked to rate or comment on the Intelligent Level Crossing, as the in-vehicle interface was not considered feasible for use by motorcyclists.

Motorcyclists' responses to the Comprehensive Risk Control crossing were generally positive, although respondents also criticised some aspects of the design. Specifically, respondents raised concerns about the safety and cost of the attendants, and having cyclists positioned at the front of the traffic queue when the lights turned green. The most commonly mentioned positive feature was the yellow road markings, although some noted that road paint can cause adhesion issues for motorcyclists, particularly in wet conditions.

Responses to the Community Courtyard crossing were generally negative, with several respondents indicating that there was nothing that they liked about the design. Respondents were especially concerned about the lack of boom barriers and audible alerts, and disliked the slow train speed. Those who did like the design commented positively on the raised approach, the traffic lights, and the large clear zone between the stop line and the tracks.

10.4.4 Urban Cyclists

Ten respondents (2 female, 8 male) aged between 18 and 64 years completed the urban cyclists' survey, of whom one had professional experience relevant to rail level crossing safety. Most of the respondents were aged between 55 and 64 years (60%) and lived in Victoria (80%), with one respondent from Queensland and one from South Australia. Data from one additional participant were excluded because they reported technical difficulties with one of the videos.

Respondents cycled an average of 93 km per week ($SD = 67$) but crossed rail level crossings relatively infrequently, with only 20% reporting that they encountered level crossings several times a week, 50% several times a month and 30% less than a month. The sample was evenly split between those whose main reason for cycling was primarily commuting and those who cycled mainly for recreation.

Paired comparison analysis indicated that cyclists rated the Comprehensive Risk Control crossing highest for safety and compliance. In contrast, the Community Courtyard crossing was rated first for efficiency and preference, but last for safety and compliance. Cyclists were not asked to rate or comment on the Intelligent Level Crossing, as they could not use the in-vehicle interface.

Cyclists' responses to the Comprehensive Risk Control crossing were quite consistent, with most users giving similar responses in terms of what elements were liked and disliked. They liked the clear warnings, especially the yellow box markings, and the cycle box. However, they criticised the fact that the design required cyclists to merge with vehicular traffic to cross the tracks, suggesting it would be preferable to maintain a designated cycle lane throughout the crossing. Consistent with feedback from other road users, several cyclists felt the attendants were redundant.

Responses to the Community Courtyard crossing were less positive. Some liked specific aspects, such as the shared zone, raised platform and stopping far back from the train tracks. However, some felt that the shared space still prioritised motorists or that motorists would disregard the signals. Cyclists expressed concern that the roadway through the crossing was too narrow, requiring them to merge with vehicular traffic, and suggested again that a separate bicycle lane would be preferable. Others commented on the lack of standard rail level crossing controls, such as barriers and audible warnings. One respondent suggested that, although the design might not work in the Australian context, it could work in other jurisdictions (i.e. with different norms and modal share).

10.4.5   Urban Pedestrians

Nineteen respondents (12 female, 7 male) aged between 18 and 64 years completed the urban pedestrians' survey, of whom five had professional experience relevant to rail level crossing safety. Over two-thirds of the respondents were aged between 25 and 44 years (68%). Most participants lived in Victoria (89%) with the remainder residing in Queensland (11%).

There was considerable variability in exposure to rail level crossings: 32% encountered them less than a month, 16% several times a month, 32% several times a week and 22% daily or several times a day. There was also variability in modal use, ranging from 5 to 70 km per week. Most respondents reported that they walked mainly for recreation (47%) or commuting (37%).

Paired comparison analysis indicated that pedestrians rated the Comprehensive Risk Control crossing highest for safety, compliance and preference, but last for efficiency. Conversely, the Community Courtyard crossing was ranked first for efficiency, but last for safety and compliance.

Pedestrians generally had very positive responses to the Comprehensive Risk Control crossing, with several stating that they did not dislike anything about the design. Features that were highlighted as being liked included the pedestrian-specific features, namely the pedestrian shelter, ticket machine and stop/go lights on the pedestrian gates. As with other road user groups, some pedestrians commented that the attendants were unnecessary and expressed concern about the cost to implement the design.

For the Intelligent Level Crossing, the design feature relevant to pedestrians was a dynamic visual display located on approach to the pedestrian gates, which provided real-time information on approaching trains. Nearly all pedestrians commented positively on the visual display, noting that the extra information was useful, as it provided information about when their train would arrive and how long they would have to wait at the crossing. Some respondents suggested this could reduce violations, because users would have a better understanding of the situation. For instance, it would deter people from running across the tracks (e.g. because they mistakenly believe the approaching train is the one they are intending to catch) and would also reduce the likelihood of people assuming the signals are in a 'fail safe' state (i.e. where the warnings are activated due to a technical fault, but no train is approaching). However, some respondents noted that the information could also encourage violations, if people tried to run in front of the train (e.g. knowing that it is their train approaching), or if they became over-reliant on the system information and did not make their own visual checks for trains. Finally, some respondents felt that the display was too close to the road and wanted greater separation between vehicular and pedestrian traffic.

Pedestrians' responses to the Community Courtyard crossing were mostly negative, with several respondents stating that there was nothing that they liked about the design.

The primary criticism was that it felt unsafe, due to the lack of physical barriers separating pedestrians and trains. Some respondents stated that shared space areas were unsafe for pedestrians. Respondents also felt that the crossing warnings were not sufficiently conspicuous and that people might accidentally walk onto the tracks when a train was approaching. As with other road users, the features that were liked related to the aesthetic, open nature of the space. Pedestrians also liked the fact that it gave them greater visibility of approaching trains and that cars would be stopped farther back at the traffic lights.

10.4.6 Summary of Responses to Urban Designs

The top-ranked urban designs for each road user group and criterion are shown in Table 10.2. This shows some clear trends: all road user groups ranked the

TABLE 10.2
Summary of Urban Road Users' Top-Ranked Designs by Criterion

Road User	Safety	Compliance	Efficiency	Preference
	Comprehensive Risk Control	Comprehensive Risk Control	Intelligent Level Crossing	Intelligent Level Crossing
	Comprehensive Risk Control	Comprehensive Risk Control	Comprehensive Risk Control	Comprehensive Risk Control
	Comprehensive Risk Control	Comprehensive Risk Control	Standard active	Standard active and Comprehensive Risk Control
	Comprehensive Risk Control	Comprehensive Risk Control	Community Courtyard	Community Courtyard
	Comprehensive Risk Control	Comprehensive Risk Control	Community Courtyard	Comprehensive Risk Control

Note: Motorcyclists and cyclists were not asked to rate the Intelligent Level Crossing as this design did not include any features intended for these road user groups.

Comprehensive Risk Control crossing first for both safety and compliance. This aligns with the design philosophy that inspired the Comprehensive Risk Control crossing, which was to include multiple redundant controls to improve safety and increase compliance among all road user groups. For heavy vehicle drivers, the Comprehensive Risk Control crossing was also the top-rated design for both efficiency and preference. In contrast, drivers rated the Intelligent Level Crossing first for preference and efficiency, whereas cyclists ranked the Community Courtyard crossing highest on these criteria.

Notably, across all road user groups except pedestrians, preference was strongly correlated with efficiency. This finding reinforces the need for system designers to focus on both safety and efficiency, rather than treating them as competing priorities, to create systems that will be acceptable to users.

10.5 RURAL DESIGN EVALUATIONS

10.5.1 RURAL CAR DRIVERS

Thirteen respondents (4 female, 9 male) aged from 18 years to more than 75 years old years completed the rural car drivers' survey. Of these, four had professional experience relevant to rail level crossing safety: two had 5–6 years' experience and two had 35–40 years' experience.

Over two-thirds of respondents held a full/open driver's licence (69%) and just over half of the respondents were aged between 45 and 64 years (54%). Most respondents lived in Victoria (46%) or New South Wales (38%), with one from each of Queensland and Western Australia. There was variability in participants' exposure to rail level crossings, with 15% reporting that they crossed level crossings several times a day, 39% several times a week, 23% several times a month and 3% less than once a month. Respondents reported driving an average of 368 km each week ($SD = 268$), with most nominating driving at work (54%) as the primary purpose for most of their driving.

Paired comparison analysis revealed that car drivers rated the Simple But Strong crossing first for compliance and preference, and second on the other criteria. The GPS Average Speed interface was ranked first for efficiency, and the standard active crossing was ranked first for safety. The standard passive crossing was ranked last on all the criteria.

Rural drivers liked most aspects of the Simple But Strong crossing, especially the advance warning and the coloured road markings. However, several respondents felt that the 40 km/h speed limit through the crossing was unnecessary, especially when no train was approaching; one driver suggested that the speed limit could be a variable message sign activated by the train. Respondents also noted that the design put the onus on drivers to monitor their speed and the environment (e.g. in comparison with boom barriers, which would 'force' them to stop) and were concerned that inattentive country drivers may miss the signs.

Drivers' responses to the Ecological Interface Design crossing were mixed. Respondents liked the coloured road markings and felt these would be effective at alerting drivers to the rail level crossing and the need to change their behaviour. Some liked the mirrors, but others were concerned about how they would perform under variable weather conditions (e.g. glare, fog). The most common criticisms were the lack of active controls and regulatory signage to guide road user behaviour. Some respondents also objected to having the train speed reduced to 20 km/h, with one respondent suggesting that this could negatively impact safety by encouraging drivers to try to 'beat the train'.

Responses to the GPS Average Speed interface were also mixed. Most drivers thought the idea had merit, but also expressed concerns about its implementation. Specific concerns were: Drivers may use the speed guidance to increase their speed and race in front of a train, GPS technology is not sufficiently reliable to implement the system, drivers may become over-reliant on technology and auditory alerts could be annoying or distracting.

Interestingly, among the drivers with professional experience relating to rail level crossing safety, there appeared to be a generational divide. The respondents with 35–40 years' experience considered the GPS Average Speed interface 'promising' but were strongly opposed to both other novel designs. In contrast, the respondents with 5–6 years' experience had a more tempered view of the GPS interface, raising both benefits and drawbacks (e.g. drivers may try to 'race' the train), and were more positive about the other novel designs, especially the Simple But Strong crossing.

10.5.2 Rural Heavy Vehicle Drivers

Five male professional drivers completed the rural heavy vehicle drivers' survey. Four were aged between 25 and 54 years, reportedly drove >4,000 km per week and encountered rail level crossings at least several times a week. The remaining driver was aged 65–74 years, reported driving 600 km per week and encountered rail level crossings with less-than-monthly frequency. Two were from Victoria, two from South Australia and one from Queensland.

Paired comparison analysis revealed that heavy vehicle drivers rated the GPS Average Speed interface highest for compliance, efficiency and preference, whereas the Simple But Strong crossing was ranked first for safety. As with car drivers, the standard passive crossing was ranked last on all the criteria.

All but one driver liked the Simple But Strong design, with the one objector noting he would prefer road markings rather than signs. One driver suggested that, although he liked the design, the reduced speed limits were problematic as many drivers disregard existing 80 km/h limits on similar roads.

Similarly, all but one driver liked the Ecological Interface Design crossing; the driver who disliked it suggested that it was not an improvement over existing designs (the only novel design that this particular driver liked was the Simple But Strong crossing). The other drivers liked the visual markings of the Ecological Interface Design crossing, and two suggested that the mirrors could be particularly useful for crossings where the road and tracks are at unusual angles. However, two of the drivers questioned the expense of installing or maintaining the mirrors, with one suggesting that vandalism could be a problem.

Heavy vehicle drivers were divided regarding the GPS Average Speed interface, with three drivers very enthusiastic about it and two opposed to it. Drivers expressed concerns about cost and feasibility, including those who regarded it positively. For instance, one driver noted that successful implementation would require coordination between rail and road transport authorities, which could be challenging. Another driver suggested the system would need (in their view, unrealistically) detailed information such as train length, to be useful.

10.5.3 Rural Motorcyclists

Forty-nine respondents (1 female, 48 male) aged between 18 and 74 years completed the rural motorcyclists' survey, of whom one had professional experience relevant to rail level crossing safety. Over half of the respondents were aged between 45 and 64 years (55%), and most held a full/open motorcycle licence (86%). Respondents lived in Queensland (37%), Victoria (31%), New South Wales (16%), South Australia (8%) and Western Australia (8%).

Respondents reported riding an average of 292 km each week ($SD = 194$), with most nominating recreation (61%) as the main purpose for most of their riding. The sample were infrequent users of rail level crossings, with 55% reporting that they encountered level crossings with less-than-monthly frequency, 31% several times a month, 8% several times a week and only 6% daily or several times a day.

Paired comparison analysis revealed that motorcyclists rated the standard active crossing first for safety and preference, and ranked the standard passive crossing first for efficiency (but last on all other criteria). The Simple But Strong crossing was ranked first for compliance. Motorcyclists were not asked to rate or comment on the GPS Average Speed interface, as the system was not designed for two-wheelers.

Motorcyclists' responses to the Simple But Strong crossing were mixed: most liked at least some aspects of the design (with some liking all of it and disliking nothing), but several stated that they did not like anything about it. Respondents generally commented positively on the active warnings, advance warning signs and 'danger zone' markings. However, many riders disliked the speed limit reduction and expressed concern that 'painted' road markings could create a hazard for two-wheelers. Some motorcyclists also suggested that inattentive riders may fail to notice the crossing until it is too late and more markings on the road would be useful. Several respondents suggested that boom barriers or grade separation would be preferable.

Responses were similarly mixed for the Ecological Interface Design crossing. Most riders liked the coloured road markings and felt that these were particularly attention grabbing, although as with the Simple But Strong design, they noted that if the colour was achieved through road paint, this could be hazardous for two-wheelers. They also liked the train's salience and reduced speed, although some expressed concern that this could encourage road users to try and race in front of an oncoming train. A few respondents liked the mirrors, but many expressed concern about the impact of sun glare. Several respondents disliked the lack of active warnings and boom barriers, and one suggested grade separation instead. A criticism unique to motorcyclists was the presence of roadside poles, which some riders thought could obscure signage and create a hazard if a motorcyclist needed to swerve in an emergency (e.g. to avoid an animal on the road).

10.5.4 Rural Cyclists

Five respondents (2 female, 3 male) aged between 18 and 54 years completed the rural cyclists' survey. Most were aged between 35 and 54 years (80%), and none had any professional experience related to rail level crossing safety. Two respondents lived in Victoria, one in New South Wales, one in Queensland and one in Western Australia.

Respondents cycled an average of 214 km per week ($SD = 156$) but crossed rail level crossings infrequently, with most reporting that they encountered level crossings less than once a month. All but one rider stated that their primary reason for cycling was recreation.

Paired comparison analysis revealed that cyclists rated the rural designs consistently on all the criteria. Specifically, the Simple But Strong crossing was ranked first on all the four criteria, the standard active crossing second, the Ecological Interface Design third and the standard passive crossing last.

Having ranked the Simple But Strong crossing highest on all of the evaluation criteria, comments on this design were generally positive with riders particularly liking the active warnings and the road markings. Two respondents indicated that there was nothing that they disliked about the design. The remainder suggested adding warnings on the road surface, widening the 'danger zone' so it extended farther back from the tracks and adding boom barriers.

Responses to the Ecological Interface Design crossing were more mixed. Cyclists liked the coloured road markings, which raise awareness and give advance warning of the crossing. The warning aspect of the road markings may be more relevant to cyclists, who travel at slower speeds than motorists. However, several cyclists disliked the mirrors, expressing concerns about the sun glare, cost and potential for vandalism. One cyclist disliked the lack of boom barriers.

10.5.5 SUMMARY OF RESPONSES TO RURAL DESIGNS

The top-ranked rural designs for each road user group by criterion are shown in Table 10.3. Whereas evaluations of the urban designs revealed consistent trends across all road users, responses to the rural designs were considerably more variable. All road users ranked either Simple But Strong or the standard active crossing highest for safety: both designs feature flashing lights and audible warnings at the crossing. Consistent with this, several road users expressed dislike of the lack of active warnings in other designs. The Simple But Strong crossing was also the top-rated crossing for compliance by all road users except heavy vehicle drivers and was the number one preference for car drivers and cyclists. This pattern of results starkly contrasts with the findings from the driving simulator studies, where Simple But Strong was clearly the least preferred design (see Chapter 9). This suggests that further evaluation is needed to understand why different evaluations produce conflicting results. One possible explanation is that when physically driving the simulation, participants were more attentive to the environment and the need to slow down, whereas in the online videos, it was easy to miss the speed limit changes. Notably, not all respondents mentioned the speed limit reduction, but those who did mention it provided negative criticism (i.e. they disliked it or thought others would not comply).

Heavy vehicle drivers ranked the GPS Average Speed interface first on three of the four criteria, which supports the original design intention for restriction of this

TABLE 10.3
Summary of Rural Road Users' Top-Ranked Designs by Criterion

Road User	Safety	Compliance	Efficiency	Preference
	Standard active	Simple But Strong	GPS Average Speed	Simple But Strong
	Simple But Strong	GPS Average Speed	GPS Average Speed	GPS Average Speed
	Standard active	Simple But Strong	Standard Passive	Standard active
	Simple But Strong	Simple But Strong	Simple But Strong	Simple But Strong

Note: Motorcyclists and cyclists were not asked to rate the GPS Average Speed design
as it used an in-vehicle interface that was not designed for two-wheelers.

concept to heavy vehicles only. However, participants raised valid concerns regarding the feasibility of implementing this system and its real-world reliability.

A common theme throughout the responses to the open-ended questions was that road users were supportive of the inclusion of coloured road markings to delineate the crossing. Indeed, several respondents (especially motorcyclists and cyclists) suggested that road markings were more useful than roadside signs. Respondents noted that country roads are monotonous, which can facilitate inattention, and therefore it is necessary to provide warnings that effectively highlight the presence of a rail level crossing. Among the designs evaluated, the Ecological Interface Design crossing appeared to be most effective in this regard as the coloured zone started well in advance of the crossing, minimising the chance that drivers and riders would fail to notice it until it was too late to stop. However, respondents disliked other aspects of the Ecological Interface Design, particularly the lack of active warnings, and expressed concern about sun glare reflecting off the mirrors.

10.6 SUMMARY

This chapter described an online survey study that elicited evaluations of proposed rail level crossing designs from different road user groups in geographically diverse areas. To our knowledge, this form of evaluation has not been widely used previously. Although the format has potential as a more cost-effective alternative to simulator studies, it also had some limitations. For the urban designs, the results for drivers were broadly consistent with the driving simulator studies in which participants experienced the rail level crossings directly. For the rural designs, however, the simulator and survey evaluations yielded conflicting results. Here it is worth noting that the novel rural designs were more revolutionary than the novel urban designs, as the urban designs largely employed controls and features that were familiar to road users (albeit not in the level crossing context). In contrast, the novel rural designs incorporated new road markings and signage, and many survey respondents objected to this lack of familiarity.

Additionally, differences in both participant demographics and the participant experience could account for discrepancies between the survey and simulator evaluations of rural designs. Potentially, the lack of immersion in the environment led to key features being missed or interpreted differently in the survey study. Conversely, participants responding to the survey study may have taken more time to consider each design, as they were provided with a description of the design and the opportunity to pause video recordings and replay them to explore points of interest in more detail. Unfortunately, we were unable to record the time taken to complete the survey or the approaches taken by respondents to determine if they were using more intuitive or rational decision-making strategies. Further research using similar approaches might consider using methodologies that would enable such data to be collected.

In relation to recruitment and sample sizes, this study was advertised widely via social media, which reaches thousands of potential participants, but response rates are typically extremely low. Although online research increases the range of potential participants, in practice response rates are low because participants are not invested in the research (e.g. they do not know the research team or the university and do not perceive any tangible benefits from participating). It may be easier to recruit participants for studies where the participant inclusion criteria are either very broad (e.g. car drivers in urban areas) or appeal to specific interest groups (e.g. rural motorcyclists). In this study, it appears that our survey benefitted from snowball sampling among rural motorcyclists, resulting in motorcyclists being relatively over-represented among rural road users. Overall, these factors mean that although the overall sample was larger than that of our simulator studies, some of the groups had very low response rates and biased samples.

Although the study had some limitations, it provided a relatively low-cost means to identify which design concepts and features were more and less preferred by a diverse group of rail level crossing users. It identified some consistent points of feedback across road users, which provide general community perceptions about issues such as government investment in design features (e.g. level crossing attendants and in-vehicle interfaces).

Importantly, the responses from road users often aligned with practical considerations that had already been raised during the design refinement process involving rail level crossing stakeholder representatives. For example, the risk to two-wheelers of painted road surfaces had been considered, and the refined concept was to use coloured road aggregate rather than paint to mitigate this potential hazard (a level of detail we were unable to commute to respondents due to the brevity of the survey).

This indicates that the stakeholders had been successful in representing the views and requirements of diverse rail level crossing users and incorporating appropriate modifications to designs.

Interestingly, the practical concerns raised by respondents around the implementation of designs, such as cost to governments (and therefore to tax payers), reliability of technology and safety of staff (i.e. crossing attendants) suggest that acceptance by end users is not limited to concerns regarding their personal interactions with the crossing. This emphasises the role of users as intelligent agents, in line with the sociotechnical systems theory perspective. That is, road users are not akin to machines that simply receive information (i.e. warnings), interpret the information and take subsequent action; instead, they come to the task of crossing the railway line with different needs, requirements, experiences, knowledge and preferences. It is important to understand these diverse views as part of the design life cycle process to avoid recommending designs that, despite receiving promising results from objective evaluations, fail to be implemented or accepted due to poor user or wider community support.

Finally, the results of the survey reinforce the finding from Phases 1 and 2 of the research programme that different road user groups have different requirements at rail level crossings. Even our design process, which explicitly aimed to design for individual differences, experienced difficulties balancing these needs across diverse groups. However, it should be noted that some designs, particularly the Comprehensive Risk Control crossing, appeared to achieve this balance quite well and arguably more effectively than existing designs.

Section V

Conclusions and Future Applications

11 Summary and Conclusions

11.1 INTRODUCTION

Safety at rail level crossings continues to be a longstanding issue, which has proven resistant to existing interventions. The research programme described throughout this book was underpinned by the notion that both safety and efficiency gains can be made through adopting a systems thinking approach. The intention was to apply systems thinking analysis and design methodologies to create new rail level crossing design concepts that, when implemented, would provide a safer environment for all users.

In this chapter, we begin by outlining some of the key strengths and contributions of the approach adopted. Following this, recommendations arising from the research programme are discussed. We then return to the questions that we posed at the very beginning of this journey. Firstly, we consider what we learnt about the factors across road and rail systems that influence user behaviour at rail level crossings, and contemplate whether the systems approach provided insights that would not have occurred through applying stand-alone human factors methods. Next, we consider whether our design process, based on the principles of sociotechnical systems theory, generated designs that jointly optimised the social and technological aspects of rail level crossing systems. Where joint optimisation was not achieved, we consider why the designs fell short of this goal. This leads to a general discussion highlighting the key lessons learned from undertaking sociotechnical systems-based design. Finally, we identify the opportunities to extend this research programme.

11.2 A WHOLE OF LIFE CYCLE HUMAN FACTORS APPROACH

An important feature of the research programme was the use of human factors analysis and design approaches throughout the rail level crossing design life cycle. The majority of the research focussed on the early aspects of the life cycle; that is, methods such as Cognitive Work Analysis (CWA) and Hierarchical Task Analysis (HTA) were used to analyse existing rail level crossing environments (Chapter 5), to inform the design of new rail level crossing environments (Chapter 6) and to assess and refine the resulting rail level crossing design concepts (Chapters 7 and 8). Importantly, within these activities, issues across the system life cycle (i.e. implementation, maintenance, upgrades, decommissioning) were considered. As one example, engagement with stakeholders during design refinement activities enabled the information to be gathered regarding maintenance costs and concerns in relation to new rail level crossing technologies and infrastructure.

This whole of life cycle human factors approach is something that is often urged in system design, but is not often achieved (Stanton et al. 2013b). As many issues emerge from unexpected interactions at the boundary between people and systems, the need to engage in an iterative design–test–redesign process is paramount. This research programme has demonstrated how this can be achieved with approaches such as CWA.

As discussed earlier, an increasing number of researchers are arguing for a systems approach to be taken when attempting to improve transportation safety (Cornelissen et al. 2015, Larsson et al. 2010, McClure et al. 2015, Salmon et al. 2012b, Salmon and Lenné 2015). As research in the transportation domain has predominantly adopted an engineering-based approach, it is important to clarify the contribution of human factors and systems thinking approaches over and above the engineering approach to facilitate further systems thinking applications.

In applying the whole of life cycle approach, the current research programme demonstrated the benefits of applying systems thinking approaches to the analysis and design of level crossing systems. By collecting comprehensive data regarding the behaviour of different users and integrating these data within appropriate systems analysis frameworks, a rich and detailed description of rail level crossing system behaviour was produced. This considered multiple forms of users, along with various systemic factors that influence behaviour. Notably, these factors included those related to multiple stakeholders: from road users, rail level crossing designers, road and rail operators, to government and the community at large. A major strength of the approach is that it promotes consideration of not only the physical environment of the crossing but also the wider environment within which rail level crossings are designed, operated, maintained and upgraded. As such, recommendations arising from the research addressed changes at higher levels of the system (e.g. risk assessment processes, incident reporting systems, design standards and guidelines) in addition to changes to the design of rail level crossing environments (e.g. warnings, signage). The breadth of the analyses conducted throughout the research programme is such that they could also inform other reforms designed to increase the reliability, efficiency and usability of rail level crossings.

Another notable strength of the approach adopted is the range of human factors concepts considered. Through applying in-depth analysis methods such as CWA, HTA and Systematic Human Error Reduction and Prediction Approach, various aspects of behaviour were considered, including decision-making, situation awareness, errors and constraints. This aspect of systems analysis methods offers a significant advantage over methodologies that focus explicitly on individual concepts (e.g. human error) in isolation.

Finally, the formative component of the analysis and design approach is worth mentioning. This enabled analysts to explore how activity within the system could be undertaken given design modifications, which in turn supported the identification of important design insights. Using standard normative or descriptive analysis approaches would not have facilitated this.

Some pertinent weaknesses of the approach should also be noted. The analysis and design process adopted incurred a significant level of resource usage, with the overall programme of research taking 5 years to complete. This could be improved through the introduction of dedicated software support for some of the analysis methods (e.g. the strategies analysis diagram) and the use of automated data collection and analysis techniques (e.g. auto-transcription). Overall, however, the utility of the outputs produced in this case justify the high level of resources invested.

11.3   RECOMMENDATIONS FOR IMPROVING RAIL LEVEL CROSSING SAFETY

The systems approach adopted ensured that various recommendations for improving safety at rail level crossings emerged. The recommendations are summarised using Rasmussen's (1997) risk management framework in Figure 11.1.

The recommendations presented in Figure 11.1 can be broadly categorised as follows:

1. Recommendations for the development of in-vehicle devices
2. Recommendations for changes to infrastructure at urban rail level crossings
3. Recommendations for changes to infrastructure at rural rail level crossings
4. Recommendations for improving safety management around rail level crossings generally

Importantly, although a number of recommendations are focussed on changing the physical environment of crossings, the research programme produced recommendations that span the entire system.

11.4   REFLECTIONS ON THE RESEARCH PROGRAMME

11.4.1   FACTORS INFLUENCING USER BEHAVIOUR

In Phase 1 of the project, described in Chapter 4, we used human factors methods to collect a wide range of data about human behaviour at rail level crossings. This represented a significant research effort in its own right and highlighted some important findings. For example, we found differences in how novice versus experienced drivers interact with rail level crossings, with novices expecting active controls at all crossings (an expectancy violated at passive crossings) and failing to consider the possibility of queuing at busy urban crossings. Further, the findings highlighted differences in information use across different types of road user.

However, using these data to generate the systems descriptions of rail level crossing with CWA and HTA provided insights over and above those generated from the initial analyses. For example, the Work Domain Analysis (WDA) presented in Chapter 5 provided an actor- and event-independent description of rail level crossing environments, identifying the various physical and abstract constraints on their functioning. This provided crucial insights into the wider systemic issues that drive the configuration of rail level crossing environments, and thus influence road user behaviour. The systemic factors identified via the WDA included the underlying assumption of the system that road users must give way to trains, the competing

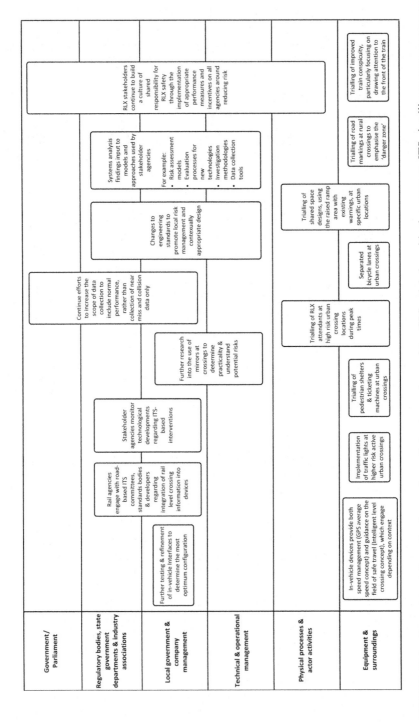

FIGURE 11.1 Recommendations for improving rail level crossing safety. GPS, global positioning system; ITS, intelligent transport system; RLX, rail level crossing.

pressures around safety and network efficiency and the focus on conformity with standards in system design and redesign.

It is worth noting that, for the research team, the WDA provided an in-depth overview of the functioning of rail level crossing systems and a way to identify potential conflicts and issues across the system. The WDA representation also resonated well with stakeholders as it provided a tangible way to see the how elements in the system were connected, in a way not previously available.

The relationship between the environment and the behaviour was then emphasised through application of the strategies analysis diagram, which explicitly links behaviours to the physical features identified in the WDA. This provided insights into how the underlying structure of the rail level crossing system ultimately influences user behaviour.

Related to this, another important outcome of the systems modelling was a detailed understanding of how different types of road users make decisions to stop or proceed (decision ladders) and the strategies they use to traverse crossings (strategies analysis diagram). The systems analyses enabled us to clearly identify where different road user groups adopt similar or consistent strategies, as well as where conflicts may arise. In short, we found considerable variation in how road users interact with rail level crossings and a range of ways in which adverse events can emerge. This raised implications for both design and redesign; for example, it was vital to be cognisant that design solutions aimed at car drivers would not have the same impact on heavy vehicle drivers or pedestrians. This was again emphasised in the differential ratings of novel rail level crossing designs provided by different road user types (Chapter 10). The importance of considering all road users in analysis and design is thus emphasised, as is the need to continually monitor and assess system behaviour and the unique challenges faced by different users.

Finally, the contextual activity template (CAT) and the Social Organisation and Cooperation Analysis (SOCA) prompted us to take very different perspectives on the system, compared with traditional human factors analyses that focus on tasks and activities. The CAT revealed the situational constraints on how the system can function and highlighted a range of possibilities for improving system design through modifications to constraints. For example, the CAT showed that information about risk is not currently provided to road users as they approach and traverse rail level crossings. This raised the idea of providing this information on approach to individual crossings, rather than holding it in documents and databases not accessible to users. The SOCA allowed us to examine the roles that human and non-human actors currently play in rail level crossing system operation, along with the roles that they could potentially play given design modifications. This enabled identification of insights that supported the generation of new design ideas. For example, when examining the function 'Alert user to presence of rail level crossing', the SOCA identified that actors outside of the road and rail infrastructure could potentially fulfil this function. This included the potential for using an in-vehicle display to warn of an upcoming rail level crossing. This potential was realised in both the Intelligent Level Crossing and GPS Average Speed design concepts.

Overall, the argument that systems analysis provides considerable insight beyond traditional human factors studies has implications for the design of future human factors research programmes. Given the acceptance within safety science that transportation systems are complex sociotechnical systems, this raises the question of how we can better support research that is integrated within a wider systems thinking framework. Acknowledging that many research endeavours do not have the time, resources or buy-in from funding agencies required to undertake in-depth systems analysis, there is a need to encourage future applications where systems models can be shared across research teams. This can avoid reinforcing system design recommendations that cater for some user types to the detriment of others, or that introduce the components into the system that have unintended and unforeseen negative consequences. For example, a comprehensive systems model of a road transport system could be shared by research groups and used to determine the design of smaller studies with the findings 'rolled up' to continually refine the larger model. Such efforts may be facilitated by the current trends towards open science and new technologies for collaboration.

11.4.2 Joint Optimisation of Rail Level Crossing Systems

In addition to the need to better understand the factors influencing user behaviour, we argued in Chapter 1 that the existing rail level crossing designs do not represent a jointly optimised system. Broadly speaking, there are three distinct design philosophies: technology-centric designs, human-centric designs and jointly optimised designs. *Technology-centric designs* are characterised by the introduction of new technologies where human activities are required to adapt to the technology. Conversely, *human-centric designs* are characterised by the introduction of new working practices, for which technologies are brought into support. In contrast, *jointly optimised designs* occur where design is used to enable successful performance to emerge from the *interaction* between human and technological aspects of the system. Essentially, the human and technological systems are considered together and are designed to be synergistic.

Existing rail level crossing environments are typically technology-centric. Here, the interaction fundamentally relies upon road users becoming aware of warnings provided to signal an upcoming level crossing (passive crossing) or an approaching train (active crossing) and complying with legislated requirements to stop and give way to approaching trains. This is a rigid system, based heavily in technology (both the physical sense and the abstract sense in the form of rules), that leaves little room for flexibility or adaptation. Indeed, even the existing process for re-designing level crossings is technology-focussed, through the application of engineering standards. The only existing intervention to the authors' knowledge that could be considered a social intervention is education and awareness campaigns, but even here the focus is typically on educating users about the rules governing their behaviour at crossings, not about positive social outcome for users themselves.

The Cognitive Work Analysis Design Toolkit (CWA-DT) design process changed this focus. The process itself was socially based – a participatory group process where stakeholders and subject-matter experts shared knowledge and expertise and worked collaboratively to develop novel ideas. This broadened the design space beyond the application of engineering standards to explore new opportunities for system functioning. In addition, several tools used within the design process were specifically focussed on priming participants to consider the social aspects of rail level crossing functioning. For example, scenarios and personas were intended to better engender a sense of empathy with different types of road users. In addition, the consideration of sociotechnical systems theory values and how they can be applied within rail level crossing design was intended to re-focus design efforts on the needs of humans within the system, including opportunities for designing transport systems that support the quality of life of their users.

This sociotechnical systems theory-inspired design process generated several designs (or aspects of designs) that were more human-centric in nature. These included the following:

- The removal of existing warning technologies and replacement with human supervisors in the Community Courtyard crossing.
- The use of a shared space road design on approach to the Community Courtyard crossing that prioritises active transport.
- Slowing of road vehicles and/or trains in the Community Courtyard, Simple But Strong and Ecological Interface Design crossing concepts, which was intended to provide vehicle operators, especially train drivers, with more control in emergency situations.
- The support for social interaction and knowledge sharing intended by the inclusion of cafes and community hubs in the Community Courtyard crossing.
- The scheduling of regular forums in regional areas to facilitate discussions between truck drivers, train drivers and system stakeholders about safety issues at level crossings and build empathy, as part of the GPS Average Speed concept.

When considering the six novel rail level crossing concepts as whole designs, it is possible to categorise them as either technology-centric, human-centric or jointly optimised designs. This is represented in Figure 11.2, which shows the extent to which the different designs produced in this research programme align with the three design philosophies.

As shown in Figure 11.2, the Ecological Interface Design crossing is the concept best aligned with the jointly optimised design philosophy. The Ecological Interface Design crossing uses minimal technology and attempts to optimise the interaction between road users and trains by clarifying or amplifying the important constraints such as the train, drivers' speed and the proximity of the train to the crossing. The Comprehensive Risk Control crossing, conversely, is the most technology-centric concept, as it uses various forms of technology to attempt to control road user behaviour. This includes boom gates, in-road LED lights, pedestrian gates and adaptive bollards. By comparison, the Community Courtyard crossing, shown at the bottom of Figure 11.2, is the most human-centric of the designs. The Community Courtyard crossing does not use active warnings at the crossing; instead, it implements a shared space concept whereby human interactions are relied upon for safe performance.

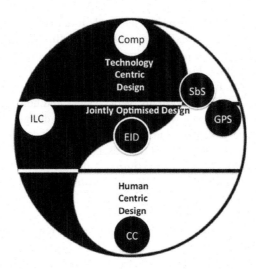

FIGURE 11.2 The alignment of rail level crossing design concepts with design philosophies. CC, Community Courtyard crossing; Comp, Comprehensive Risk Control crossing; EID, Ecological Interface Design crossing; GPS, GPS Average Speed interface; ILC, Intelligent Level Crossing; SbS, Simple But Strong crossing.

11.4.3 Shifting Paradigms

Sociotechnical systems theory represents a very different paradigm compared with the technology-centric safety and risk management approach currently applied in rail level crossing design (see Chapter 6). Two fundamental issues that limit the application of sociotechnical systems theory are the extensive resources required such as time, cost, expertise (Clegg 2000) and stakeholder acceptance. Within the current project, we introduced sociotechnical systems theory to stakeholders through 3 days of workshops, which was potentially insufficient time given the inherent conflicts between sociotechnical systems theory and existing design philosophies. Our participatory approach involved efforts by system stakeholders to reconcile the two approaches, by combining the protective aspects of the safety management approach while incorporating some of the sociotechnical systems theory values and principles. Given the sensitivity of the stakeholders to the political realities of transportation system design, this appears to be a practical and reasonable approach. However, it limits the promised benefits of sociotechnical systems theory in promoting flexible and adaptable systems.

It is reasonable that genuine paradigm shifts would need to occur over a much longer period of time. The use of the CWA-DT in this context can be viewed as the beginning of an ongoing process to introduce the sociotechnical systems theory approach and integrate it into rail level crossing design and evaluation processes. The workshops provided a means to initiate conversations about systems thinking and sociotechnical systems theory among a diverse group of stakeholders from road and rail domains, including engineers, policy officers, human factors researchers, consultants and safety executives. Potentially, this experience might have positive effects on how the participants approach safety issues in their future work, while also strengthening professional networks across the various stakeholder groups.

The two expert-generated designs – the Community Courtyard and Ecological Interface Design crossings – offered more radical change in the functioning of rail level crossing systems. Both designs involved slowing the train through the crossing and did not feature any standard regulatory signage and warnings such as flashing lights. The Ecological Interface Design crossing concept demonstrated positive features in the evaluation process; however, further research is needed to determine whether all aspects of the design (such as mirrors) would be effective in the real world. The Community Courtyard crossing, however, was received quite poorly in user evaluations, with all road users considering it to be unsafe and motorists ranking it as least efficient. Although non-motorised road users rated it higher for efficiency, its overall acceptance by users remains questionable.

The poor response to the Community Courtyard crossing is again indicative of key tensions between sociotechnical systems theory and the traditional safety management approach usually applied in road design. For example, the sociotechnical systems theory value of humans as assets suggests that humans in the system should be given control over their decisions, while the principle of flexible specification (i.e. that design should only specify that which is necessary) advocates that humans should be supported to exhibit flexibility and adaptability in their behaviour. In contrast, the safety management approach encompasses concepts such as the hierarchy of control, which are focussed on separating humans from hazards. When applied within transport systems, these concepts tend to focus on limiting performance variability rather than supporting adaptive variability. Therefore, designs based on sociotechnical systems theory are likely to violate road users' expectations about what is acceptable in road design, due to their deviation from established conventions.

Further obstacles to embracing sociotechnical systems theory-based designs include law reform and public opinion. Although many international jurisdictions have adopted a 'safe system' approach to road safety (Salmon et al. 2013b), none have yet attempted to deal with the law reform issues that would be required to facilitate implementation of a strategy that genuinely embraces shared responsibility. To truly support sociotechnical systems theory approaches to design in public safety domains, many changes to the 'protective structure' (Dekker 2011) would be required. For example, the operation of the legal system and the interventions of regulators need to be considered in relation to what organisational behaviour is rewarded and punished.

Moreover, a general trend in public opinion away from individual responsibility for safety towards government responsibility has been noted (Leveson 2004). This shift is evident in cases of civil proceedings against organisations and governments, where it is contended that governments hold a duty of care towards the public. In the rail level crossing context, it appears anecdotally that if active warnings are in place where a collision occurs and the equipment operated as designed (i.e. provided a warning), the public will attribute the incident to the road user as they have breached the road rules. However, if the equipment failed to warn of a train, the railway company and/ or government are deemed responsible. This situation has stifled the implementation of low-cost innovations at rail level crossings (Road Safety Committee 2008).

Focussing on culpability in accident investigations also limits the data collected and, consequently, our ability to learn from past mistakes.

Although have advocated for the sociotechnical systems theory approach in this research programme, the question remains: can we be assured that designs based on this approach would be more successful in preventing accidents than traditional approaches? To answer this important question further research is required to compare the traditional safety management approach with sociotechnical systems theory, to determine which is most successful in preventing accidents. Potentially, modelling approaches such as agent-based modelling and systems dynamics (see Section 11.5.2) could provide answers.

11.4.4   Reflections on Sociotechnical Systems Theory-Based Design

As emphasised throughout this research programme, the translation of analysis outputs and findings into design has been a longstanding challenge for CWA (Jenkins et al. 2010, Lintern 2005, Mendoza et al. 2011, Read et al. 2015a, Stanton et al. 2013a), and indeed for human factors more generally (Dul et al. 2012). Although there have been successful design applications in the past, including using approaches such as Ecological Interface Design (McIlroy and Stanton 2015), the lack of guidance for using CWA in design more broadly has potentially affected its usability, accessibility and its uptake in practice. The development and application of the CWA-DT within this research programme intended to fill an important gap in CWA practice and has the potential to improve the translation of the systems approach in real-world design. Although this was achieved in some respects, as the CWA-DT could generate the novel design concepts presented in this book, it is worth discussing the extent to which stakeholders valued the new process.

The CWA-DT process was well received by stakeholders, with positive subjective ratings of the process gained through our evaluations (see Chapter 8). Stakeholders and subject-matter experts found it beneficial to think creatively and to work collaboratively with individuals from across the system with different views and perspectives. In addition, the process was successful in generating four novel design concepts, which were taken forward for evaluation and testing. However, the design process did raise some interesting questions and implications for future applications.

Firstly, there is some debate within human factors about the role of users in design. Eason (1991) discusses the knowledge-into-use approach, where users constitute the sources of data for design and may be involved as participants in requirements-gathering activities and user testing of prototypes. It has been noted that, in many human-centred design processes, the user is spoken for by the researcher who collects and synthesises their data (Sanders 2002). Eason (1991) contrasts this with the user participation approach, adopted in sociotechnical systems theory-based design, where the user is the client and has decision-making power in the design process. Proponents of the former approach argue that users may not be the best judge of their needs and require experts to make design decisions in their best interest. Supporters of the participatory approach, conversely, argue that users should determine how values and objectives are traded-off in design as they retain ongoing ownership of the system, whereas designers will move on to other systems.

In our research programme, we adopted a hybrid approach beginning with participatory design and supplementing this with an expert design process. Further research could investigate whether a collaborative process, involving both stakeholders and researchers throughout the analysis and design process, would achieve superior results.

11.5 FURTHER RESEARCH OPPORTUNITIES TO IMPROVE RAIL LEVEL CROSSING SAFETY

The research programme led to the identification of several future research areas that could build on the work undertaken. These include field trials of proposed design changes to rail level crossing environments, the application of computer simulation and modelling techniques to explore the impact of design changes, and research efforts to better integrate the findings of human factors research into economic modelling for cost–benefit analysis.

11.5.1 FIELD TRIALS

Firstly, there are opportunities for additional evaluation processes to support the potential uptake of recommended design changes. One potential means could be the testing of prototype designs in controlled field trials. For example, a prototype of the mirrors from the Ecological Interface Design concept could be built and tested via controlled field trials. Such trials could initially involve the use of test tracks, with participant observers in stationary vehicles as trains traverse (to mitigate the risk of participants being involved in collisions if there is a poor response to the mirrors), under various environmental conditions. This could be done with a range of road user types, including car drivers, heavy vehicle drivers, motorcyclists and cyclists, and could enable collection of data around reaction times, situation awareness, decision-making and the potential for emergent behaviours introduced by the new design.

Other interventions could potentially be implemented and tested in the real world according to the notion of safe-to-fail experiments proposed by Snowden and Boone (2007). This approach acknowledges that, when implementing changes in complex systems, we are unable to predict the effects until they are implemented in practice. Following the safe-to-fail approach, design concepts could be introduced in stages, beginning with the least novel components (i.e. for the Ecological Interface Design concept – adding guide posts to create an illusion of travelling at higher speed) and ending with the most radical (i.e. the replacement of traditional signage with mirrors). With careful monitoring of behaviour occurring during the transition, it would be possible to incrementally measure the impacts on behaviour, including the identification of emergent behaviours not initially predicted. This enables a change in strategy to occur quickly if monitoring reveals potential risks. Such an approach to evaluation requires a flexible and ongoing commitment to evaluation and thoughtful planning to ensure its appropriateness when conducted in high-hazard domains. Further, it requires a culture that accepts failure and easily adapts to try new innovations, which is unusual in safety-critical industries due to strict requirements for demonstrating the effectiveness of safety interventions prior to implementation. This again highlights where the current protective structure may impede the application of systems thinking in transportation design.

11.5.2 System Simulation and Modelling

Where real-world testing proves either too expensive or too risky, the use of system simulation and modelling can provide a useful alternative. For example, methods such as systems dynamics (Sterman 2000) and agent-based modelling have previously been used to simulate the impacts of changes within transportation systems (e.g. Goh and Love 2012, McClure et al. 2015, Shepherd 2014).

Agent-based modelling is a computer modelling approach in which rules are applied to direct the behaviour of individual agents who then interact, enabling the identification of emergent properties arising from the interactions.

An agent-based model typically comprises the following three elements (Macal and North 2010):

1. A set of agents, each with attributes and behaviours
2. A set of agent relationships and methods of interaction
3. The agents' environment

This enables the analyst to model both, interactions between agents and interactions between agents and their environment. The modelling therefore provides a bottom-up analysis of system functioning and enables analysts to identify how changes to the rules governing interactions lead to emergent behaviours and affect the system as a whole. Agent-based modelling techniques have previously been used to model transportation systems (Sarvi and Kuwahara 2007, Thompson et al. 2015).

Systems dynamics, another computation modelling technique, is generally used to understand the implications of wider system change, such as policy change. Models are usually represented in diagrammatical form such as stock and flow diagrams with the relationships between factors in the model underpinned by differential equations (Hettinger et al. 2015). Both agent-based modelling and systems dynamics have limitations, and thus adaptations or hybrid approaches have been proposed to address these issues (e.g. Hettinger et al. 2015, Thompson et al. 2015). Importantly, it is proposed that these types of modelling techniques could be directly informed by HTA and CWA outputs. For example, the values and priority measures in an abstraction hierarchy could become variables in systems dynamics models, whereas the strategies identified in the Strategies Analysis phase can help to identify the behaviours of agents well as the rules or conditions under which those behaviours would be more likely to occur.

Of course, the validity of any modelling approaches must be addressed, and the outcomes will rely on the quality of the inputs and the assumptions made by analysts. Yet, these approaches provide a promising way to evaluate and compare options, and to identify potential issues not previously recognised.

A strength of computational modelling approaches is that they analyse system functioning over time. The approaches taken in the current research programme described rail level crossing systems at a point in time, but given that road and rail networks are dynamic open systems, these models may not remain accurate for long. The increasing uptake of new technologies in transportation systems, including higher forms of automation in both road vehicles and trains, raises questions about how system functioning will change in future. Furthermore, it is unknown how the recommended rail level crossing design changes might interact with a more highly automated transport system. Modelling techniques may assist to test assumptions about future systems and how interactions might change over time.

11.5.3 Cost–Benefit Analysis

A final area for further research in rail level crossing safety relates to approaches to cost–benefit analysis, with cost often raised as a key constraint by stakeholders. Benefit–cost ratios for upgrading crossings to existing standards for active warnings are low (Wullems et al. 2013), and it was clear from the Kerang accident discussed in Chapter 1 that budgetary constraints are a key reason why rail level crossing safety is difficult to address.

Cost–benefit information is required by stakeholders across the system to inform decision-making, from decisions about research and development trials, to adoption of design changes or new technologies, to changes to policy and regulation at the government level. Approaches to cost–benefit analysis tend to allocate a cost value to the prevention of fatalities or to other units of cost, such as disability-adjusted life years, so that a ratio can be determined. Although human factors researchers should not necessarily be responsible for conducting comprehensive cost–benefit analyses, we do have a role in providing information that can be used in these calculations, or working with economists and other professionals to ensure that evidence of benefits is appropriately integrated into these approaches.

Importantly, the outcomes of human factors analyses need to be integrated into safety risk models. This is a fruitful area for further research, and it is suggested that modelling approaches, such as systems dynamics and agent-based modelling, informed by systems analyses and human performance data, may provide the important gains in this area.

In addition to finding better ways to integrate findings from human factors studies into risk and economic analyses, human factors and sociotechnical systems theory approaches may be able to provide additional benefit by promoting a systems view on benefits beyond lives lost and disability. If there are additional outcomes expected by implementation of a design – such as productivity, improved quality of life for individuals or increased social capital within a community – then potentially systems models can begin to place a monetary value on these and demonstrate the wider benefits of jointly optimising transportation systems.

11.6 SUMMARY

The culmination of the research programme provides a watershed moment within transportation safety research as it represents the first attempt at implementing systems thinking and sociotechnical systems theory throughout the transport design life cycle. By sharing our approach and learning, we intend to support others to conduct similar efforts to address other complex problems in transport and beyond. Chapter 12 will outline some potential applications and initial case studies of further applications.

12 Future Applications and Opportunities

12.1 INTRODUCTION

Although the research programme described in this book was focussed on reducing trauma at rail level crossings, the approach we adopted is generic in nature and as such it lends itself to applications in many other domains.

In this chapter, we provide some suggested directions for future research applications that would benefit from adopting similar approaches to those described in this book. We have demonstrated that our approach was beneficial in its application to rail level crossings, but there are myriad opportunities to conduct similar applications to solve other issues facing complex systems, both in transportation and beyond. Among these new research directions, we also provide some examples where we have begun to apply these approaches, particularly Cognitive Work Analysis (CWA) and the CWA Design Toolkit (CWA-DT), to new areas.

12.2 FURTHER APPLICATIONS IN TRANSPORTATION SYSTEMS

All transportation domains exhibit the properties of complex systems and can benefit from the application of systems thinking approaches. These applications relate to both the physical design of transport systems and higher level regulatory and policy design. In relation to road transport, for example, the design of road environments would benefit from the approach adopted in this book, as would the design of regulatory and policy frameworks. Accordingly, we describe a further application of the CWA-DT to create intersections that better support compatible situation awareness across road user types. We also discuss potential directions for research into the human factors challenges associated with the introduction of autonomous vehicles. Following this, we explore opportunities for further research in the rail, maritime and aviation domains.

12.2.1 IMPROVING INTERSECTION DESIGN

Collisions at intersections represent a major road safety issue. In Australia, for example, the majority of urban crashes and a substantial proportion of rural crashes occur at intersections (McLean et al. 2010). As with rail level crossings, the problem is not limited to a single road user group, with most forms of road user having elevated crash risk at intersections. In the Australian state of Victoria, for example, during 2012 over half of all car, bicycle, motorcycle and pedestrian crashes occurred at intersections (VicRoads 2013).

As with rail level crossings, the prevalent approach to understanding and preventing collisions at intersections has tended to be component driven, focussing on specific road user types (e.g. drivers) and/or fixing a component of the problem (e.g. increasing the conspicuity of motorcycles; Gershon and Shinar 2013). Previous studies of behaviour at intersections, for example, have largely been road user centric, focussing on individual road users (e.g. drivers, cyclists, motorcyclists or pedestrians alone) or factors such as driving errors (Gstalter and Fastenmeier 2010, Sandin 2009), pedestrian behaviours (King et al. 2009), and engineering (e.g. Highways Agency 2012). The authors have recently completed a research programme in which the intersection crash problem was tackled using a systems approach (see Box 12.1).

12.2.2 Responding to the Challenge of Highly Automated Vehicles

An important emerging challenge for human factors is the introduction of highly autonomous vehicles into road networks. This could potentially represent a step-change in the operation of the road transport system, with considerable anticipated benefits for road safety such as reduced crashes due to the prevention or improved tolerance of 'human error'. However, the shift toward full automation is occurring within already complex and poorly understood road systems, with fatalities on a scale comparable to major public health issues such as cancer, and cardiovascular and respiratory diseases (Salmon and Lenné 2015, World Health Organisation 2014). The period between now and fully automated driving will see new emergent driving behaviours, new interactions between drivers and technologies and ultimately new forms of road traffic crashes. Key potential issues include poor intervention by human supervisors (due to vigilance decrements and skill degradation over time), design or software coding failures that are opaque and lie latent and the potential for software hacking leading to widespread disruption of road networks. Although the technology is racing ahead, our understanding of advanced automated systems and their likely impacts on behaviour is not keeping pace (Banks et al. 2014). In addition, appropriate policy frameworks have not yet been fully developed. This could lead to the design and implementation of systems that will not be fit for purpose (Banks and Stanton 2016), along with policy-related issues such as confusion around liability following crashes involving automated vehicles.

It is critical therefore that designers of automated road transport systems adopt sophisticated approaches that have the capacity to consider emergent risks. Here, systems models can respond to the challenge by providing input into the design of vehicles, infrastructure, policy and regulation surrounding this new way of operating. Approaches such as CWA and the CWA-DT could play a key role in automated road system design. In particular, the formative capacity of CWA would enable consideration of emergent risks. For example, based on an in-depth formative analysis of different levels of automation in road transport systems, the CWA-DT could be used to inform the following areas with the aim of achieving a jointly optimised automated driving system:

- Guidance on policy requirements for automated driving systems.
- Optimal allocation of functions between drivers and technologies (i.e. which automation technologies should do what, which driving functions humans should retain).

BOX 12.1   DESIGNING NOVEL INTERSECTIONS

As part of a research programme that aimed to improve situation awareness on the road, we applied the CWA-DT following a systems analysis of road intersections (see Salmon et al. 2014a,b). The aim of the design process was to generate new intersections to better support compatible situation awareness among different types of road users. Initially, naturalistic data were collected from drivers, cyclists, motorcyclists and pedestrians as they traversed an urban road network, including a range of intersections, while providing verbal protocols. These data were used to build an understanding of the system and interactions between road users using Neisser's (1976) perceptual cycle model (Salmon et al. 2014a), the Event Analysis of Systemic Teamwork (EAST) methodology (Salmon et al. 2014b) and the Work Domain Analysis phase of CWA.

Insights from these analyses were used to develop materials for participatory design workshops involving 11 subject-matter experts employed in academia, industry and government. Participants' disciplinary backgrounds included human factors, psychology, sociology, traffic engineering, urban planning and safety science. Here, we adapted the approach used for rail level crossings, by integrating the research team with the other design participants to facilitate better multidisciplinary learning and debate within the design process. Within the participant group, there were experienced users representing a range of road user types (e.g. drivers, pedestrians, cyclists, a motorcyclist and a heavy vehicle driver).

Participants received a presentation providing an overview of the research problem, key findings from the systems analyses and an introduction to the sociotechnical systems theory values and principles. They then undertook exercises from the CWA-DT to promote creativity. These included lateral thinking exercises, assumption crushing, metaphors and use of the Design with Intent cards (Lockton et al. 2010; see Chapter 6 for a description of these tools). In addition, we added a new design tool – constraint crushing – where design participants reviewed the Work Domain Analysis, identified key constraints and discussed whether constraints could be:

- Made visible to users through design.
- Strengthened to further restrict behaviour.
- Removed to enable more flexible behaviour.

This exercise enabled the participants to become familiar with the Work Domain Analysis output and prompted better consideration of constraints in design. Further details about the design process are available in another publication (Read et al. 2015b).

(Continued)

BOX 12.1 (Continued) DESIGNING NOVEL INTERSECTIONS

Participants were asked to redesign a particular intersection that was identified as poorly designed during the initial data collection phase. The process produced three intersection concepts (see Figure 12.1):

- *Turning teams*: This concept bases priority on road user type and direction of travel, with users joining 'teams' that move together through the intersection on designated lanes that use colour to delineate the path through the intersection. A filtering box for motorcyclists and cyclists is provided at the head of the traffic queue. Pedestrian crossing points are matched to desire lines, based on the location of buildings and adjoining pedestrian paths. The pedestrian crossing path is also widened to enable cyclists to have an official alternative of crossing with pedestrians if they are uncomfortable traversing the intersection with the motorised traffic.
- *Self-regulating intersection*: This intersection design is based on the principles of a roundabout; however, it uses a large oval-shaped median strip in the centre of the intersection. When traffic from each intersecting road is given priority to enter the intersection (via filtering traffic lights), they move around the median strip in the same direction and exit where they wish. Cyclists have the option to either move with the motorised traffic or 'cut through' via dedicated lanes available through the central median strip. Within the intersection, there are no lane markings to promote connectedness between road users and require them to negotiate their way through with other road users at slow speeds.
- *Circular concept*: This concept is based on better separation between motorised and non-motorised traffic. Pedestrian crossing zones are provided farther back from the intersection than usual, but footpaths are linked in a circular pathway, that is, shared with cyclists turning left or right. This circular pathway links with cycle lanes placed in the centre of the intersecting roads. Quality of life for the local community is enhanced through converting existing unused space around the intersection to parkland with outdoor cafes, gardens, barbecue areas and seating provided.

This application is an example where the CWA-DT process has been applied to develop innovative designs in another domain, aimed at jointly optimising the interactions between humans and technology. The initial designs have been subject to a design refinement process but further testing and evaluation, using methods such as driving simulation and field trials, are the next important steps to fulfil the whole of life cycle CWA-DT approach.

(Continued)

BOX 12.1 (Continued)   DESIGNING NOVEL INTERSECTIONS

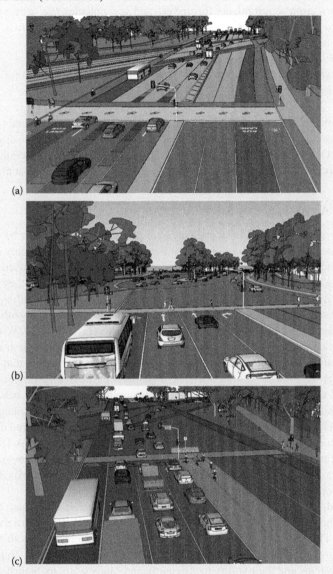

FIGURE 12.1   Novel intersection design concepts. (a) Turning teams intersection design concept, (b) Self-regulating intersection design concept and (c) Circular intersection design concept.

- Proposed interfaces for key automation system displays (visual, auditory, haptic and multi-modal).
- Information requirements for drivers and technologies (i.e. who needs to know what, how situation awareness should be distributed across drivers and technologies).
- Standard operating procedures for use.
- Optimal integration of vehicle, automation and infrastructure-based systems.

Further, integration of methods such as CWA with computational modelling approaches (e.g. systems dynamics and agent-based modelling; discussed in Chapter 11) could be used to proactively identify potential risks associated with the introduction of automated vehicles into the existing road transport system.

12.2.3 Additional Rail Safety Applications

Although accidents at rail level crossings are generally considered to be one of the biggest risks to safety in the rail industry, there are a range of other safety issues that could be addressed through these methods. For example, signals passed at danger represent a key breakdown in rail safety systems (Naweed 2013), as they can result in more than one train in the same section of track simultaneously, leading to a collision, or a train entering a section that is otherwise unsafe, resulting in collisions with track workers, derailments or other adverse outcomes (ITSR 2011). CWA has been previously applied in rail to design improved interfaces for automatic train control systems using a user-centred design process with train driver subject-matter experts (Jansson et al. 2006). Although a less structured participatory process than the CWA-DT was used, it followed a similar philosophy, with the users being encouraged to find a solution that could best meet their needs and could be integrated into their ways of working. It would be interesting to compare a user-centred design process, such as that undertaken by Jansson et al. (2006), with the CWA-DT process, to understand any differences in process (i.e. the types of issues discussed and how decisions are made) and design outcomes (i.e. which approach is more likely to result in jointly optimised system design system).

Beyond safety applications, systems thinking approaches could also be used to improve the efficiency of the rail network by informing work design to network management. This could leverage from earlier CWA applications such as those designing improved interfaces for air traffic controllers (Ahlstrom 2005). Reliability of railway systems and their ability to efficiently and safely adapt to disturbances, such as incidents (including level crossing collisions) and other delays, is an important area for human factors research, given the role that the rail network plays in meeting mass transit and freight needs (Golightly and Dadashi 2017). Here, a systems approach can assist by identifying how system constraints hinder flexibility and how they might be modified to provide controllers with more options. Additionally, systems approaches could be used to improve the way information about constraints is made available to controllers to assist them to take action, particularly during abnormal and emergency situations.

12.2.4 MARITIME

The approach to transport analysis and design adopted for our rail level crossing research could also be usefully applied within the maritime domain. Perrow (1999) describes maritime transport as an 'error-inducing system'. A diverse industry, maritime operations occur in harsh environmental conditions, with strong commercial interests leading to strong production pressures and a focus on liability.

Given the complexity of the maritime transport system, the need for systems thinking in maritime has been recognised in previous research and systems thinking applications. For example, Rasmussen's AcciMap approach has been used to analyse large-scale ferry disasters, including the Zeebrugge Herald of Free Enterprise (Svedung and Rasmussen 2000) and the more recent South Korea Sewol ferry accident (Kee et al. 2017). These analyses have demonstrated how systemic factors in the maritime industry have interacted to lead to these serious events. Interestingly, however, to our knowledge, CWA has not previously been used to evaluate or contribute to the re-design of ferry operations. It is suggested that such work could assist in efforts to improve safety.

In military maritime operations, however, CWA has been applied to understand the work domain of navy frigates (Burns et al. 2005) and to analyse control room activities in the Trafalgar class submarine (Stanton 2014, Stanton and Bessell 2014; see Box 12.2).

BOX 12.2   SYSTEMS ANALYSES OF SUBMARINE OPERATIONS

CWA has been used to describe the control room of the Trafalgar class submarine. The analysis identified the system functions and explored the way in which those functions were used in situations, decisions, strategies, role allocations and required competencies (Stanton and Bessell 2014). Insights from the application of CWA indicated that the nature of these functional constraints needs to be brought into question when considering the design of future systems. For example, new systems could incorporate changes to:

- The nature of the functions being performed.
- The type of decisions made and informational cues used (to make decisions simpler and more transparent, and introduce task prompts to reduce ambiguity and memory load).
- Function allocation (such as semi-automating or fully automating some functions and/or changing role allocations).
- Competency requirements (to simplify training and work demands).

In addition to CWA, the EAST framework was used to describe the control room in terms of task, social and information networks, as well as exploring the relationships between those networks (Stanton 2014). Combined task and

(Continued)

**BOX 12.2 (Continued)   SYSTEMS ANALYSES
OF SUBMARINE OPERATIONS**

social networks provided insights into which roles were performing tasks in series and in parallel. Combined information and social networks gave insights into which roles were communicating which information concepts. Finally, integration of the three networks then provided insight into how information was used and communicated by people working together in the pursuit of tasks.

The insights gained from these analyses could be used in combination with insights from the CWA in a CWA-DT process (similar to that conducted for intersection design in Box 12.1) to inform the design of new control room environments. Importantly, the task, social and information networks could be used to evaluate the impacts of proposed design changes on the existing structure of the work through a desktop evaluation process similar to that described in Chapter 7.

12.2.5   AVIATION

The formative capacity of the approach adopted in this research programme lends itself to the design of first-of-a-kind systems. Modelling future systems is a very useful extension to methods in human factors and has been demonstrated through prior CWA applications in aviation. For example, Naikar et al. (2003) used CWA to explore possibilities for crewing on first-of-a-kind airborne early warning and control systems. This involved the use of an abstraction hierarchy and contextual activity template to evaluate potential team concepts (e.g. the number and role of team members required to operate the system) using a desktop analysis approach based on realistic scenarios. This process enabled the identification of design requirements and the development of a team concept that met these requirements.

In another application of CWA to explore design configurations in future aviation systems, Stanton et al. (2016) explored crew configurations to enable single-pilot operations (see Box 12.3).

BOX 12.3   FUTURE FLIGHT DECKS

CWA was used to explore the design of future flight decks given the predicted moves to single-pilot aviation operations (Stanton et al. 2016). This application demonstrated how such an aircraft could be ready for service entry within a decade by using a distributed air/ground sociotechnical system, rather than focussing on equipping an aircraft with complex automation to aid the pilot.

(Continued)

BOX 12.3 (Continued)   FUTURE FLIGHT DECKS

To achieve this insight, four alternative crewing options were modelled using CWA, with the models then analysed using social network analysis. The study compared the following configurations:

- Single-pilot aircraft.
- Single-pilot aircraft with additional automation mirror on the ground that cross-checks inputs and outputs independently of aircraft automation.
- Single-pilot aircraft with an additional pilot on the ground who can be called upon if needed.
- Single-pilot aircraft with both the additional automation mirror on the ground and the additional pilot on the ground.

CWA, especially the Social Organisation and Cooperation Analysis phase, which utilised the contextual activity template tool, proved useful as a formative approach to help understand the function allocation distribution between the various actors involved (e.g. the pilot flying, pilot monitoring, aircraft automation) and the functional loading on them. Extending this analysis to understand communication interactions through social network analysis revealed that the two options comprising an additional pilot on the ground would be the most resilient, potentially even more so than the current dual-pilot cockpit arrangement (Stanton et al. 2016).

This application demonstrated the utility of CWA as a way to evaluate potential design configurations, similar to the evaluations described in Chapter 7. Further, the use of social network analysis in conjunction with CWA demonstrates another useful means for desktop evaluation of design concepts that may be useful in future research applications.

12.3   APPLICATIONS BEYOND TRANSPORT

Naturally, beyond transportation, there are many domains and topics in which systems thinking in analysis, design and evaluation can provide the important gains in safety and performance. We discuss current and potential applications in a sample of diverse domains, including outdoor education, cybersecurity and sport.

12.3.1   Preventing Incidents in Outdoor Education

The systems thinking approach lends itself to the design of accident prevention strategies. As part of a wider programme of research to understand the systemic contributory factors to incidents in outdoor education, and assist the industry to take action to prevent incidents (see Salmon et al. 2017), we applied an adapted version of the CWA-DT to design injury prevention strategies (see Box 12.4 and Goode et al. 2016). This domain involves the delivery of high-risk activities in dynamic environments and has experienced multi-fatality incidents (Salmon et al. 2012a). To address the risk in this domain, the 'UPLOADS' incident reporting system was developed based on Rasmussen's (1997) risk management framework and the AcciMap technique.

BOX 12.4   PREVENTING INCIDENTS IN THE GREAT OUTDOORS

Data reported through the UPLOADS incident reporting system covering a 12-month period was used as the basis of the design process. AcciMap was used to analyse the systemic contributory factors involved in over 350 injury incidents, 86 illness incidents and 74 near-miss incidents. Notably, this included contributory factors from across all levels of the system (from environment and activity level to regulatory bodies and government level), along with the relationships between the factors.

A total of 30 participants attended two design workshops, held in two Australian cities. Participants represented actors across the sector, including those employed by secondary schools, outdoor education providers, outdoor training organisations, outdoor sector peak bodies, work health and safety regulators and relevant government departments.

Materials from the CWA-DT – such as the design brief and design criteria documents – were used to plan the workshop. In addition, creativity exercises from the CWA-DT were used during the workshops. Similar to the intersection design process described in Box 12.1, design participants were provided with the outputs of the analysis process (i.e. AcciMap representations displaying an aggregate summary of the contributory factors identified from incident reports separately for injuries, illnesses and near misses).

Participants were asked to interpret the Accimaps to identify insights, which they would then discuss in small groups, and propose strategies to address the network of contributory factors. These accident prevention strategies were documented by facilitators on a 'PreventiMap' framework (similar to the AcciMap framework, but incorporating prevention strategies to address an identified issue). Participants were encouraged to document a network of prevention strategies, with prompts used to encourage consideration of what changes at the higher levels of the system would be required to support prevention measures on the ground. For example, a change in organisational policy would be required to support the introduction of an improved risk assessment process. This process generated PreventiMaps for various issues, including the prevention and management of activity leader fatigue, the prevention of burns during cooking activities and improvement of the competency of activity leaders in conducting dynamic risk assessments. Importantly, the PreventiMaps were evaluated against Rasmussen's tenets and were found to align well with the majority of the tenets, such as incorporating strategies for actors across the various levels of the outdoor education system. However, evaluation of the process suggested it could be improved by prompting participants to identify strategies that could address the migration of work practices and erosion of risk controls over time.

12.3.2 ENHANCING APPROACHES TO CYBERSECURITY

Cybersecurity represents a topical 'wicked problem' that could be tackled through a systems approach, such as that adopted in this research programme. One cybersecurity issue that is currently problematic is identity theft and trading in the dark net. The dark net is a layer of the Internet not accessible through popular search engines such as Google. Features of the dark net include websites, discussion forums and marketplaces that trade in both legitimate and illicit products and services. As such, the dark net facilitates various criminal activities, one of which is identity theft. The buying and selling of identity credentials (e.g. passports, driving licenses, credit card details, bank details, utility bills) via dark net forums has been identified as a significant and growing global problem (Ablon et al. 2014) and one for which appropriate interventions have not yet been developed.

An effective response by individuals, organisations and law enforcement agencies depends on a holistic understanding of how dark net marketplaces operate, along with the provision of frameworks to support implementation of appropriate intervention strategies and assessment of their impact. Despite the emergence of identity theft as a global cybersecurity issue, this knowledge and framework does not currently exist (Lacey and Salmon 2015). Accordingly, the authors are currently engaged in a research programme in which systems analysis approaches such as CWA and EAST are being applied to analyse dark net trading in order to inform the design of interventions through the CWA-DT. Further applications of these approaches to other cybersecurity issues are encouraged.

12.3.3 OPTIMISING SPORTS SYSTEMS

A burgeoning area for human factors and systems thinking applications is that of sport. There are considerable similarities between sporting domains and other complex domains in which these approaches are more traditionally applied such as operating under time pressure, system dynamism and strong reliance on teamwork for successful performance. Salmon et al. (2010) discuss the potential utility of applying human factors and systems thinking analysis and design approaches in sports contexts.

More recently, a number of applications have been emerging, with analyses of sports systems (Salmon, Clacy and Dallat 2017), injury issues such as concussion identification and management (Clacy et al. 2016) and situation awareness, decision-making and teamwork (Neville et al. 2016). Sport represents an interesting application area as the potential applications cover a diverse set of sports and purposes. Possible applications span team design, injury prevention, performance analysis and enhancement, training and coaching design and regulatory framework design. Wider systemic sports issues such as doping, corruption and sports governance could also be tacked. In addition, increasing participation in sports and recreational activities contributes to wider benefits of such as active and socially cohesive communities, and as such, research in this area could inform wider systems analyses using approaches such as systems dynamics.

12.4 SUMMARY

This book described a novel research programme applied to an intractable transport problem that combined human factors and systems thinking methods to analysis, design and evaluation. The work led to the generation of valuable insights and recommendations aimed at improving safety and overall system performance. We propose that it provides a valuable model for future research.

In this chapter, we have discussed a selection of opportunities for future research that draws upon the strengths of human factors and systems thinking to address the important issues facing modern societies. We have also provided examples where this research is underway and is beginning to demonstrate positive outcomes. There is no doubt that there are many more domains and issues that could benefit from these approaches than those discussed. We hope that others will continue to build upon these ideas and to push the boundaries of human factors and systems thinking to make further in-roads in solving these important real-world problems.

Appendix: Guidance for Using the Key Human Factors Methods and Approaches

This appendix provides procedural guidance for applying the human factors methods described and used within this book. An overview of each of the methods is presented in Chapter 2. This appendix provides step-by-step guidance on how to apply each method. There are

We have also attached a new figure - labeled A.1):
"Four main types of methods will be discussed:

- Data collection methods for understanding human behaviour and performance (see Section A.2).
- Data collection methods for understanding performance of the overall system (see Section A.3).
- System-focused analysis methods (see Section A.4).
- Human factors design methods (see Section A.4).

Figure A.1 illustrates the relationships between the methods."

A.1   GENERAL CONSIDERATIONS

A.1.1   HUMAN RESEARCH ETHICS

In using methods that require the collection of data from human participants, researchers should be cognisant of, and comply with, human research ethics requirements. In Australia, research being conducted at universities must comply with the National Statement on Ethical Conduct in Human Research (2007/2015). Similar requirements exist in many other countries. Compliance with ethical standards often requires submitting a formal written application to a Human Research Ethics Committee or Institutional Review Board. Ethics approval may also be required by other organisations or agencies involved in research (e.g. employee unions, educational providers). Where ethics approval is not formally required, those undertaking research should still ensure they are aware of the standards and ensure they treat participants in an ethical manner.

Some key ethical issues that may arise from human factors research include the following:

- Ensuring that participation in research is voluntary and not coerced (especially where participants are employees).

- Ensuring the confidentiality of data collected (especially where it is sensitive or may reveal illegal activity).
- Ensuring that research does not place participants at risk of physical, psychological, social, legal or economic harm, beyond what they might experience in their everyday lives.

The guidance provided in Section A.2 is based on the assumption that ethical approval for the research has been granted by an appropriate research ethics committee. Further information about research ethics can be found in the Australian National Statement on Ethical Conduct in Human Research (2007/2015).

A.1.2   DEFINING THE AIMS AND OBJECTIVES OF THE RESEARCH

Prior to selecting and applying the methods, the research team should consider the aims and objectives of the research as this will influence the selection of appropriate methods and inform how the methods are applied. In particular, it is suggested that consideration is given to:

- What problem is being addressed by the research?
- What research questions need to be answered?
- What human factors constructs are of interest (e.g. decision-making, situation awareness, workload)?
- Which types of participants should be targeted (e.g. drivers, cyclists, motorcyclists)? Are there sub-groups of interest (e.g. novel versus experienced users)?
- Who are the other stakeholders within the system, how should their views and expertise be captured?
- What are the key tasks undertaken in the system?
- In what environments are tasks undertaken?
- What kind of data is required for analysis?

A.2   DATA COLLECTION METHODS FOR UNDERSTANDING HUMAN PERFORMANCE

This section provides guidance for using methods that collect data from individuals to understand human performance.

A.2.1   OBSERVATION

A.2.1.1   Step 1: Define the Scenario(s) for Observation

Once the aims and objectives of the analysis have been defined, the scenario(s) to be observed should be identified. For example, consideration should be given to whether the observations should be conducted during normal operations, or if it is important to also gather data during abnormal or emergency conditions. Further, consideration should be given to whether the observation will be conducted overtly or covertly. The exact nature of the required scenario(s) should be clearly defined.

A.2.1.2   Step 2: Develop the Observation Plan

Once the aim of the analysis and the type of scenario to be observed are determined, a detailed planning of the observation should occur. Consideration should be given to what needs to be observed and how this can be undertaken. A walkthrough of the system/environment/scenario under analysis can be useful for the analyst(s) to familiarise themselves with tasks of interest in terms of time taken, location of those undertaking tasks and the general functioning of the system of interest.

Standardised recording tools should be developed and the required length of observations should be determined. Appropriate placement of video and audio recording equipment should also be determined at this stage. If relying on on-site observations with no recording undertaken, the team should consider whether a second observer is required to simultaneously record data for a proportion of time to establish reliability between observers.

A.2.1.3   Step 3: Pilot Observation

In any observational study, a pilot or practice observation is crucial. This allows the analysis team to assess any problems with the data collection, such as noise interference or limitations of the recording equipment. The quality of the data collected can also be tested and any effects of the observation upon task performance can be assessed. In addition, reliability between observers can be assessed to ensure that there is an appropriate level of agreement in what is being recorded. If substantial problems are encountered, the observation procedure or the recording tool may need to be re-designed. This piloting process should be repeated until the analyst(s) are satisfied that the quality of the data collected will be sufficient for their study requirements.

A.2.1.4   Step 4: Conduct Observation

Once the observation has been designed, the observation can proceed. Observation length and timing are dependent upon the scope and requirements of the analysis.

A.2.1.5   Step 5: Descriptive Data Analysis

Once the observation is complete, the first task to support data analysis is to create a transcript of the whole observation. This is a very time-consuming process but is critical to the analysis. The transcript should be time-stamped, with the level of detail dependent on how rapidly the variables under observation change. Depending upon the analysis requirements, the research team should analyse the data in the format required to answer the research questions, for example, calculating frequency of tasks, frequency of verbal interactions, sequence of tasks and so on.

A.2.1.6   Step 6: Further Analysis

Once the on-road initial process of transcribing and coding the observational data is complete, further analysis of the data is undertaken. Depending upon the nature of the research, observational data can be used to inform a number of different analyses, such as task analysis, human error analysis, communications analysis or Cognitive Work Analysis (CWA).

Further information about observational analysis can be found in the work of Stanton et al. (2013).

A.2.2   VEHICLE MEASURES

A.2.2.1   Step 1: Determine the Vehicle Technology Required

Driving performance and vehicle measures can be collected during an on-road study. Depending on what is being measured, there are a range of technologies that can be used. In a typical on-road study of driver behaviour, participants would drive an instrumented vehicle that is fitted with a data logging device to record vehicle speed and global positioning system (GPS) location. The data logger should be connected with the vehicle's Controller Area Network to record parameters such as interactions with the brake, accelerator and steering wheel. In addition, inward facing cameras should be used to record the participant and the car interior and outward facing cameras should record the forward view, as well as to the rear and sides of the vehicle. Depending on the aims of the study, an eye tracking device may also be used (see Section A.2.3 for further information about eye tracking).

For road users other than drivers, different technologies may be required. At a minimum, GPS position and video recording of the forward view should be obtained. There are a number of relatively inexpensive technologies available that can achieve this, including head-mounted cameras for cyclists and motorcyclists and smartwatches that collect GPS and speed information.

In an on-road driving study, participants may drive a dedicated instrumented vehicle that the research team owns or leases, which has all the required technologies pre-installed. Alternatively, participants may use their own vehicle, with the researchers fitting data loggers and other technologies to the vehicle at the beginning of the session. The latter option is usually only adopted for studies that collect naturalistic data continuously over an extended time period (e.g. 1 week to 1 year).

A.2.2.2   Step 2: Design the Route

In a quasi-experimental on-road study, all participants should complete the same route, so that behavioural data can be aggregated and/compared when participants are interacting in the same environments. When designing the route, consideration should be given to:

- The types of infrastructure that participants will encounter (e.g. signalised intersections, unsignalised intersections, roundabouts).
- The length of time it will take to complete the route. This needs to be long enough to collect data in the situations of interest but not so long as to lead to participant fatigue.
- The road scenarios under analysis (e.g. low traffic vs. high traffic).

- The complexity of the route, with a preference for simple routes that can be memorised by participants.
- The proximity of start/end points to research facilities or other suitable locations for recruiting, briefing and debriefing participants.

The route should be piloted at similar times to when data will be collected (e.g. weekday mornings, after peak times) to ensure that participants will encounter the types of situations of interest for the research.

It is also advisable to contact road authorities to determine if any works are planned on the roadway during the study period.

An alternative data collection method is to have participants drive naturalistically on their usual routes, which is referred to as a naturalistic driving study. This type of naturalistic study requires considerable resources, in terms of both time and money, as well as specialised recording equipment (i.e. data loggers are installed in participants' vehicles for an extended period of time to record their daily driving behaviour).

A.2.2.3 Step 3: Brief and Train Participants

Prior to commencing the on-road study, participants should be briefed on the aims of the study and should be provided with a map of the route, designating the start and end points of the route. Participants should be asked to memorise the route and to obtain guidance from the researcher if they have any questions about the directions.

Where the route is too complex for the participant to memorise, other strategies for providing route guidance should be used. These could include a pre-programmed navigation device or, for car drivers, having the researcher present in the vehicle to provide navigation instructions during the drive.

Depending on the aims of the study and the measures being used, participants may require training on other data collection approaches such as concurrent verbal protocols (see Section A.2.4) or workload assessment techniques (see Section A.2.6).

A.2.2.4 Step 4: Conduct Study

The participant is asked to drive/ride/walk the route while behaving as he/she would in his/her everyday life. The researcher may follow behind to ensure that the participant is completing the route correctly and to assist in case of unexpected events occurring such as taking a wrong turn, encountering road works or other changes to the road environment affecting the route. If the participant is unaccompanied on the route, it is advisable to provide him/her with a mobile phone with the researcher's phone number pre-programmed, so that the researcher can be easily contacted if assistance or navigational guidance is needed.

In some studies, data may be collected during the drive by researchers present in the vehicle. For example, errors observed during the drive and actions taken to recover from errors can be recorded on a pre-specified proforma (Young et al. 2013).

After each participant has completed the route, he/she is debriefed. It is then advisable to download the data recordings and store them on a secure data storage device.

A.2.2.5   Step 5: Analyse Data

Firstly, the videos showing the context of the drive are viewed and the data files are coded according to the types of encounters experienced by participants. For example, participant 1 might experience intersection 1 as the lead vehicle, whereas participant 2 might experience the same intersection as the following vehicle. This context affects the level of decision-making control able to be exercised by the participant. It may therefore be advisable to analyse encounters involving lead vehicles separately to encounters involving following vehicles.

Once the data have been reviewed and coded, there are a number of ways in which the data can be analysed, depending upon the aims of the research. Variables such as speed and braking can be mapped over a geographical area. For example, mean speed can be calculated at intervals on approach to an intersection and represented graphically to provide speed profiles for different types of encounters.

The video recordings can also be coded to determine variables of interest such as compliant versus non-compliant behaviour by participants, or to assess how participants interact with other road users (e.g. to conduct estimates of headways when following another vehicle).

A.2.3   EYE TRACKING

Eye tracking is a useful method for studying how individuals search their environment for visual information. Analysts can examine how observers scan a scene or visual display and which objects within the environment capture their attention. Eye tracking can be used to help understand behaviour in a range of tasks. Participants can be engaged in real tasks (e.g. driving a real car, making a cup of tea), simulated tasks (e.g. in a flight simulator) or viewing static images or videos of tasks on a computer.

A.2.3.1   Step 1: Select an Eye Tracking System

There are a range of technologies available for eye tracking. Systems can be either head-mounted or mounted on a surface in front of the participant (such as dashboard systems that can be used for in-vehicle studies).

The following should be considered when selecting an appropriate system:

- Will it be used indoors (i.e. in a laboratory) or outdoors (i.e. in an on-road study). If the latter, can it record outdoors, including handling sunlight and variations in light?
- What is the recording accuracy? Optimal accuracy is <1° error for laboratory studies, whereas systems with <4°–5° errors may be acceptable in on-road studies.
- Can it be used by participants wearing corrective spectacles?
- Is it portable? (if required to be used by a pedestrian or two-wheeled road user, in multiple vehicles, or otherwise moved between locations)
- Does it require a computer connection?
- How long can it record for in a single day? (battery life and data storage capabilities)
- Does it come with a specialised analysis software?

A.2.3.2   Step 2: Determine the Calibration Procedure

It is vital to calibrate the system to ensure the accuracy of the data collected. This calibration needs to occur with each participant. Calibration methods will depend on the technology used, but it is recommended to use at least three-point calibration and to ensure that tracking is calibrated on both the horizontal and vertical planes. In addition, conducting validation procedures during and after each trial (i.e. checking the calibration and/or running the calibration procedure again) can assist the analyst to determine whether tracking has remained accurate throughout the trial.

A.2.3.3   Step 3: Pilot Test

The eye tracking system should be piloted tested with the experimental task to ensure that accurate levels of tracking can be achieved during task performance.

A.2.3.4   Step 4: Recruit Participants

It is important to ensure that the participants recruited for the study do not have attributes that will affect the accuracy of the eye tracking. For example, if using a wearable system, these cannot generally be used by people who wear corrective spectacles. Prospective participants should be asked if contact lenses can be worn instead. In addition, with many systems, it is advisable to ask participants to avoid wearing any eye make-up during participation as this may affect the accuracy of the system.

A.2.3.5   Step 5: Brief Participants

Before commencing the experiment, participants should be briefed about the eye tracking system and how it operates. When using head-mounted systems, the system should be securely fastened in place. With some systems, particularly those that are heavy or cannot be tightly fastened, it may be necessary to advise participants to avoid making unnecessary head movements following the calibration procedure as this may change the position of the glasses in relation to the eyes and affect the accuracy of tracking.

A.2.3.6   Step 6: Collect Data

At the commencement of each trial, the system should be calibrated. As mentioned previously, it is advisable to then verify the calibration at the end of each trial. Given the size of the data files produced, at the end of each experimental session it is recommended that the files be transferred onto an external, secure data storage device to ensure that adequate space remains on the recording device or computer for subsequent participant recordings.

A.2.3.7   Step 7: Analyse Data

Many eye tracking systems will have software packages that assist to analyse the data collected. The types of events that are typically of most interest to human factors researchers are fixations (i.e. periods where gaze position is fixed in a location, e.g. for 50 ms or longer) and saccades (i.e. individual eye movements from one fixation location to another). Depending on the aims and objectives of the research, analysis might focus on items of fixation, areas of fixation (e.g. for an on-road driving study, the scene might be divided into centre road ahead, off-road areas, mirrors and in-vehicle areas – see, e.g., Young et al. [2015]) and duration of fixations. Saccades can be analysed to gain an understanding of the patterns of eye movements across a scene.

Further information about the use of eye tracking in research can be found in the work of Holmqvist et al. (2011).

A.2.4 VERBAL PROTOCOL ANALYSIS

A.2.4.1 Step 1: Define the Scenario under Analysis

Firstly, the scenario to be analysed should be determined. The scenario should relate to the aims and objectives of the research and generally involves the participant interacting with the product or system of interest. In addition, a comparative analysis can be undertaken by exploring the task undertaken with different products or systems or in different contexts. For example, in the on-road study conducted by Salmon et al. (2014a), participants provided verbal protocols while undertaking the task of negotiating different types of driving environments (e.g. intersections, arterial roads, roundabouts and shopping strips).

A.2.4.2 Step 2: Instruct/Train the Participant

Once the scenario is determined, the participant should be trained used pre-prepared materials that provide information about what is required of them during the study. Box A.1 provides an example of text that could be used in an on-road study to understand driver situation awareness.

BOX A.1   EXAMPLE PARTICIPANT TRAINING INSTRUCTIONS FOR PROVIDING CONCURRENT VERBAL PROTOCOLS

We are going to use a technique known as concurrent verbal protocol analysis to gather information regarding your situation awareness (i.e. understanding of what is going on) during the drive. This involves you 'thinking aloud' as you drive the vehicle around the route. It is important that you verbalise what you are thinking/doing mentally as you drive and not what you are physically doing. We are looking for a description of the content of your thinking and awareness while driving the route, so we want you to verbalise when you are thinking about the driving task, other road users (other traffic, pedestrians), the road environment, road infrastructure, and so on, and how it all relates to what you are doing or are about to do. It is important that verbalisations are concerned with the content or outcomes of thinking.

So, we are looking for descriptions like the following:

- I am trying to work out if the car to the left of me is going to move into my lane.
- I am checking the traffic lights/speed limit sign.
- I think that this traffic light is about to turn red so I am speeding up a little.

(Continued)

> **BOX A.1 (Continued)   EXAMPLE PARTICIPANT**
> **TRAINING INSTRUCTIONS FOR PROVIDING**
> **CONCURRENT VERBAL PROTOCOLS**
>
> - I can see that the car ahead is indicating so I know that they are about to slow down and turn left.
> - The car in front is slowing down so I need to brake.
> - I do not know what the current speed limit is, so I am guessing based on the road that it is 60 km/h.
> - I just noticed the pedestrian up ahead on the left so I am keeping an eye on him/her as I think he/she is about to cross the road.
>
> Examples of things that we do not want you to verbalise include the following:
>
> - The steering wheel is a little hard to turn.
> - I am pressing the accelerator down here.
> - I wonder what I will have for dinner.
>
> It is important that you verbalise or think aloud continuously as you drive the route.

A short demonstration given by researcher or using a pre-recorded video of another person performing a concurrent verbal protocol should be used to help participants understand what is required. However, the example should not involve an identical scenario to the one being studied, to avoid participants simply repeating the example verbalisations. A practice run may also be undertaken, with feedback given by the researcher. Alternatively, a desktop driving simulator can be used for participants to practice providing concurrent verbal protocols. The researcher should observe and provide feedback as required.

A.2.4.3   Step 3: Begin the Scenario and Record the Data

The participant should begin to perform the scenario under analysis. The whole scenario should be audio recorded by the researcher. Ideally, a video recording should also be taken to understand the context under which the participant was performing.

A.2.4.4   Step 4: Transcription

Once collected, the data should be transcribed verbatim. The transcription sheet should be designed based on the analysis aims and requirements. A spreadsheet or Microsoft Word is often used so that the transcript can include timestamps and participant verbalisations, and may also include the important contextual features relating to the interaction. This aspect of verbal protocol analysis can be time consuming and laborious but is important for the rigour of the process.

A.2.4.5   Step 5: Code Verbalisations

The written transcript should then be categorised or coded. Depending upon the requirements of the analysis, the data may be coded into one of the following five categories: words, word senses, phrases, sentences or themes. The coding scheme chosen should be based upon the aims of the research being undertaken. Walker (2005) suggests that this should involve attempting to ground the coding scheme according to some established theory or approach, such as mental workload or situation awareness. The analyst should also develop a set of written instructions for the coding scheme. These instructions should be strictly adhered to and constantly referred to during the coding process (Walker 2005). Various computer software packages are available to aid the analyst with this process, such as NVivo, General Inquirer, TextQuest and Wordstation. In addition, automated software packages such as Leximancer can be used, which use algorithms to sort the data and represent the relationships between the concepts arising within the transcripts.

A.2.4.6   Step 6: Devise Other Data Columns

Once the coding is complete, the analyst should devise any 'other' data columns. This allows the analyst to note any mitigating circumstances that may have affected the verbal transcript.

A.2.4.7   Step 7: Establish Inter- and Intra-Rater Reliability

Reliability of the coding scheme needs to be established (Walker 2005) through testing its reproducibility. That is, do independent raters achieve the same results given the same data being analysed? If reasonable levels of reliability (e.g. at least 85%–90% agreement between raters) are not achieved, the scheme may need to be modified to ensure that categories are well defined and are mutually exclusive, or the training of raters may need to be improved.

A.2.4.8   Step 8: Perform Pilot Analysis

The protocol analysis procedure should now be tested within the context of a small pilot study. This will demonstrate whether the verbal data collected is useful, whether the coding system works and whether inter- and intra-rater reliability are satisfactory. Any problems highlighted through the pilot study should be refined before the analyst applies the scheme to a full dataset.

A.2.4.9   Step 9: Analyse Structure of Verbal Protocol Data

The results from verbal protocol analysis are usually presented as frequencies of occurrence for each category of the coding scheme. For a more sophisticated analysis, the data can be analysed contingent upon events that have been noted in the 'Other data' column(s) of the worksheet, or in the light of other data that have been simultaneously collected (Walker 2005).

For further information about verbal protocol analysis, see Stanton et al. (2013).

A.2.5 Cognitive Task Analysis Interviews with the Critical Decision Method

A.2.5.1 Step 1: Prepare for Interview

When conducting a Critical Decision Method (CDM) interview, it is recommended that two researchers be present to ensure that all relevant information can be captured (Klein and Armstrong 2005). The interview should be recorded using a video recording device or an audio recording device enabling the content to be transcribed for analysis.

A.2.5.2 Step 2: Select Participants

CDM involves interviewing participants about the cognitive processes they used to make a decision during an incident or situation of relevance to the research questions. It normally focusses on non-routine incidents, such as emergency scenarios, or highly challenging incidents. However, it can also be used for routine situations; in these cases, it is recommended that the interviews be conducted immediately after the routine event, to ensure that participants are genuinely recalling the event in question and not describing an idealised or stereotypical event. The types of participants to be recruited should be identified based upon the likelihood of their experience with making decisions related to the topic of interest. Considerations should also be given to recruiting different samples of participants, to enable comparisons such as between experts and novices.

A.2.5.3 Step 3: Select the Incident to Be Analysed

If the type of situation is not already pre-defined (as it might be where CDM is used in combination with an on-road study), the interview begins with asking the interviewee to think of recent high-risk or challenging situations. These are discussed briefly to enable the researcher to determine which is likely to be of most benefit in relation to the research question(s). This situation is then selected for further discussion. The interviewee involved in the CDM analysis should be the primary decision-maker in the chosen situation, and it should have occurred recently enough for the interviewee to remember the details of their decision-making process.

A.2.5.4 Step 4: Gather and Record Account of the Situation

Next the interviewee should be asked to provide a more detailed description of the situation in question, from the beginning (i.e. hearing an alarm sounding) to the end (i.e. when the incident was classed as 'under control').

A.2.5.5 Step 5: Construct Timeline

The next step in the CDM process is to construct an accurate timeline of the situation under analysis. The aim of this is to give the researcher a clear picture of the incident and its associated events, including when each event occurred and the duration of events.

According to Klein et al. (1989), the events included in the timeline should encompass any physical events, such as alarms sounding, and also 'mental' events, such as the thoughts and perceptions of the interviewee during the situation. The construction of the timeline serves to increase the researcher's knowledge and awareness of the situation while simultaneously focussing the interviewee's attention on each event that occurred.

For instances where the researcher is present for the event in question (e.g. if the CDM is being conducted as part of an on-road study), this aim is less crucial but can still be useful to establish the interviewee's recollection of the event.

A.2.5.6   Step 6: Identify Decision Points

While constructing the timeline, the researcher should select specific decisions of interest for further analysis. Each selected decision is then probed or analysed in detail. In particular, decision points where other courses of action were available should be probed further (Klein et al. 1989).

A.2.5.7   Step 7: Probe the Selected Decision Points

Each decision point selected in step 6 is analysed further using a set of specific probes. The probes used are dependent upon the aims of the analysis and the domain in which the incident is embedded. Klein et al. (1989) summarise the probes that were used in early applications of CDM. A set of revised CDM probes was developed by O'Hare et al. (2000). These include probes relating to goal specification, cue identification, expectancy, conceptual models, the influence of uncertainty, information integration, situation awareness, situation assessment, options, decision blocking, basis of choice and analogy/generalisation.

A.2.5.8   Step 8: Analyse CDM Data

Transcriptions of the interview recordings can be subject to content analysis to identify emergent themes relating to decision-making for the situations of interest (Wong 2004). Data can also be coded according to theoretically consistent frameworks, such as Klein's (1993) recognition-primed decision model. Alternatively, data can be subject to automated analysis using software such as Leximancer. Importantly, the data can be used to populate systems analysis models such as situation awareness networks or decision ladders from CWA.

For further information about CDM, see Klein et al. (1989) and Stanton et al. (2013).

A.2.6   Assessing Workload with the NASA Task Load Index (NASA-TLX)

A.2.6.1   Step 1: Define the Tasks to Be Assessed

The selection of tasks for the workload assessment will depend upon the overall aims and objectives of the research. It will often be too time consuming to assess all possible tasks relating to a system; therefore, a selection should be made that is either most representative or most important to consider.

Conducting a Hierarchical Task Analysis (HTA see Section A.4.2) may assist to identify those tasks that are most important to assess. For example, if considering the potential effect of a novel in-vehicle assistive device on driving, an HTA of the driving task may assist to identify high-risk tasks that should be assessed. These tasks might include specific components of a typical driving route such as negotiating different types of intersections (signalised, unsignalised, roundabouts, rail level crossings), lane keeping and lane changing.

A.2.6.2   Step 2: Select and Recruit Participants

Next, it is necessary to determine if particular types of participants are required to conduct the analysis, based on the aims and objectives of the research. For example, the research may be focussed on differences in the level of workload experienced by novice and experienced drivers or young and elderly drivers. If so, the type of participants and appropriate parameters (i.e. age or licensing status) should be defined. Recruitment should then focus on only participants who match the criteria.

A.2.6.3   Step 3: Brief Participants

Before performing the task under analysis, participants should be briefed on the task to be undertaken and on the NASA-TLX procedure.

The NASA-TLX scale is composed of six sub-scales: Mental Demands, Physical Demands, Temporal Demands, Performance, Effort and Frustration. Participants rate each item on a scale from 'very low' to 'very high', or 'good' to 'poor' for the performance sub-scale (see Figure A.1 for an example proforma for the NASA-TLX)."

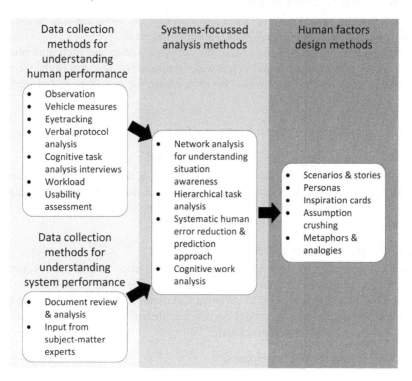

FIGURE A.1   Example proforma for the NASA-TLX.

A.2.6.4 Step 4: Performance of the Task

Participants perform the task(s) of interest. The NASA-TLX can be administered during task performance (e.g. during low workload periods of the task) or once the task is complete. It is generally recommended to administer the scale post-trial to avoid affecting the primary task performance. However, this should occur immediately post-trial so that intervening activities do not affect participants' memory of task performance.

A.2.6.5 Step 5: Administration of the Scale

Participants are presented with the scale and are asked to provide a rating between 1 and 20 for each sub-scale. Administration can be done using a pencil and paper form or using a computerised version of the scale.

A.2.6.6 Step 6: Weighting Procedure

A weighting procedure is available as part of NASA-TLX administration but is not always used. Here, participants are presented with 15 pairwise comparisons of the six sub-scale factors (Mental Demands, Physical Demands, Temporal Demands, Performance, Effort and Frustration) and are asked to select the factor from each pair that contributed most to the workload for the task. The experimenter then tallies the number of times each factor was selected by the participant in the pairwise comparisons, and this becomes the weighting for each.

A.2.6.7 Step 7: Score Calculation

If the weighting procedure is not used, the scores for the sub-scales can be used as a raw task load index for each sub-scale for further analysis.

If the weighting procedure is used, the sub-scale scores are each multiplied by the weightings calculated from the pairwise comparisons. The sum of the weighted ratings for each task is then divided by 15 to obtain the overall weighted workload score for that participant.

A.2.7 USABILITY ASSESSMENT USING THE SYSTEM USABILITY SCALE

A.2.7.1 Step 1: Define the Tasks to Be Assessed

The first step in using the System Usability Scale (SUS, Brooke 1996) is to determine the key tasks that will be assessed for usability, given that it is generally not feasible to assess all tasks. As with the NASA-TLX, an HTA can assist to identify the tasks most relevant to the aims and objectives of the study.

A.2.7.2 Step 2: Select and Recruit Participants

Next, it is necessary to determine if particular types of participants are required to conduct the analysis. For example, the research may be focussed on differences in the level of usability perceived by different end users. If so, the type of participants and appropriate parameters should be defined. Recruitment should then focus on only participants who match the criteria.

A.2.7.3 Step 3: Brief Participants

Before performing the task under analysis, participants should be briefed on the task to be undertaken and on the SUS procedure.

A.2.7.4 Step 4: Performance of the Task

Participants use the product or device, or experience the environment of interest, ensuring they complete all of the tasks defined in step 21.

A.2.7.5 Step 5: Administration of the Scale

Following task performance, participants complete the SUS questionnaire, which consists of 10 usability statements that are rated on a scale from 1 (strongly agree) to 5 (strongly disagree). The statements are as follows (Brooke 1996):

1. I think that I would like to use this system frequently.
2. I found the system unnecessarily complex.
3. I thought the system was easy to use.
4. I think that I would need the support of a technical person to be able to use this system.
5. I found the various functions in the system were well integrated.
6. I thought there was too much inconsistency in this system.
7. I would image that most people would learn to use this system very quickly.
8. I found the system very cumbersome to use.
9. I felt very confident using the system.
10. I needed to learn a lot of things before I could get going with this system.

A.2.7.6 Step 6: Score Calculation

The SUS is scored by first calculating the score for each scale item. For positive items (1, 3, 5, 7 and 9), the score contribution is the scale position minus 1. For negative items (2, 4, 6, 8 and 10), the contribution is 5 minus the scale position. The sum of the scores is then multiplied by 2.5 to obtain an overall usability value between 0 and 100, with higher scores indicating greater usability.

A.3 DATA COLLECTION METHODS FOR UNDERSTANDING SYSTEM PERFORMANCE

This section provides guidance for collecting data about system functioning that can be used to input to system-focussed analysis methods.

A.3.1 DOCUMENT REVIEW AND ANALYSIS

A.3.1.1 Step 1: Identify the Types of Documentation

Taking into account the aims and scope of the analysis, determine the types of documentation that will be of use to review. For example, when conducting a CWA of a system, the types of documents that are often used include the following:

- Engineering specifications.
- Training materials.
- Standard operating procedures.
- Government policy documents.
- Incident and/or accident reports.

Speaking to domain experts may also assist in identifying the relevant documents that can inform the analysis.

A.3.1.2   Step 2: Source Documents

Documents may be publicly available and thus sourced through Internet searches. However, other documents may only be available through domain experts, and thus, cooperation with relevant stakeholder organisations is important to gain access to such information.

A.3.1.3   Step 3: Analyse Documents

Documents should be analysed in a manner relevant to the broader research programme. For example, if using document analysis to build an abstraction hierarchy for CWA, the documents can be reviewed and key information coded (using a software program such as NVivo), based on the level of the hierarchy to which they relate.

A.3.2   INPUT FROM SUBJECT-MATTER EXPERTS

A.3.2.1   Step 1: Identify the Relevant Subject-Matter Experts

The aims and scope of the research will assist to define which subject-matter experts (SMEs) should be included. Using a tool such as Rasmussen's ActorMap (1997) can assist to ensure that all necessary roles and organisations are represented. Although it may not be possible to speak to SMEs from all stakeholder organisations, it is recommended to recruit SMEs representing each level of the system.

A.3.2.2   Step 2: Recruit SMEs

Access to domain experts or SMEs can be difficult, especially when they hold operational roles (such as pilots or train drivers) where their time is scheduled according to service delivery requirements. Therefore, the development of relationships with stakeholder organisations is important to ensure the support of the research and access to such resources.

A.3.2.3   Step 3: Develop Data Collection Plan

Data collection with SMEs could involve semi-structured interviews as well as other methods such as CDM interviews or focus groups. Protocols should be developed to ensure that questions asked are relevant to the overall research questions and the wider analyses being conducted (e.g. HTA, CWA). Questions should be piloted prior to use with SMEs to ensure that they are logical and will be easily understood by the target participants.

Walkthroughs of a system can also be a valuable way of gaining familiarisation with the system, with the SME providing information about how tasks are conducted or how the overall system operates.

A.3.2.4   Step 4: Conduct Data Collection

Interviews can be conducted in person or by telephone/Skype if required. Interviews should be audio recorded to enable later transcription and analysis.

For activities such as walkthroughs, it is beneficial for multiple research team members to attend to ensure that all relevant information can be captured through note taking and to ensure that analysts gain familiarisation with the system.

A.3.2.5 Step 5: Data Analysis

As with document review, the data should be analysed based on the wider research programme to which the SME data relates. Any recorded data should be transcribed and should be coded according to the needs of the wider analysis.

A.4 SYSTEMS-FOCUSSED ANALYSIS METHODS

This section provides guidance for conducting systems-based analysis methods.

A.4.1 NETWORK ANALYSIS FOR UNDERSTANDING SITUATION AWARENESS

A.4.1.1 Step 1: Define the Scenario under Analysis

The first step in conducting network analysis to understand situation awareness is to define the scenarios of interest, based on the overall aims of the research. These scenarios will define the type of data that are collected through methods such as verbal protocol analysis (see Section A.2.4) or cognitive task analysis interviews (see Section A.2.5).

A.4.1.2 Step 2: Data Collection

Data to inform situation awareness networks can be gained from verbal protocol analysis or CDM methods.

A.4.1.3 Step 3: Data Preparation

Participant recordings should be fully transcribed. Timestamps should be used when transcribing verbal protocol data, and the transcript should be coded accorded to key scenarios of interest and any other comparisons that the analyst intends to explore (e.g. novice–expert participants).

A.4.1.4 Step 4: Network Construction

There are two methods for constructing the situation awareness networks: manual network construction and automated network construction.

The manual network construction process involves a content analysis procedure where the transcripts are reviewed to identify and code situation awareness concepts (e.g. road sign, vehicle, speed, other road users, roadworks) and the relationships between them (e.g. vehicle 'has' speed, worksite 'has' workers). For example, from the sentence 'I am checking the worksite for workers', the concepts 'worksite', 'workers' and 'checking' would be coded. In addition, relationships between 'checking' and 'worksite', and 'workers', as well as a relationship between 'worksite' and 'workers', would be coded.

Although the manual situation awareness network construction process has more rigour and can be more sensitive to differences across participants, it is highly resource intensive. A useful alternative when there are time and resource constraints is the use of automated network construction tools such as Leximancer.

Leximancer uses text representations of natural language to interrogate verbal transcripts and identify themes, concepts and the relationships between them. The software does this by using algorithms linked to an in-built thesaurus and by focussing on features within the verbal transcripts, such as word proximity, quantity and salience. Initially, Leximancer looks for words that frequently appear in the text and then uses a weighting procedure to classify frequently appearing words as concepts. Once a list of concepts is identified, Leximancer determines how concepts are related to one another by measuring the co-occurrence of concepts within the text. Leximancer thus automates the content analysis procedure by processing verbal transcript data through five stages: conversion of raw text data, concept identification, thesaurus learning, concept location and mapping of relationships. The output is a network representing concepts derived from the verbal transcript and the relationships between them reflected within the verbalisations. Although manual construction of situation awareness networks is more sensitive to differences across participants, the Leximancer tool is especially important to analyses of this kind because it provides a less resource intensive, reliable and repeatable process for constructing situation awareness networks and removes analyst subjectivity during network creation.

A.4.1.5   Step 5: Network Analysis

Once networks are generated for the situations of interest, they can be analysed to understand their structure. Typically, a range of metrics are used to analyse situation awareness networks. These can include network density, sociometric status, cohesion, centrality and network diameter.

Analysis of data using network analysis metrics is normally supported through network analysis software tools such as Agna. This involves importing the network data into Agna in the form of a matrix of the concepts (e.g. 'car', 'traffic lights', 'pedestrian', 'speed') and the relationships between them (e.g. 'car' was mentioned with 'speed' seven times in the transcript, 'car' was mentioned with 'pedestrian' once). The network metrics are then calculated automatically by the software tool by selecting the appropriate metrics.

Metrics that may have particular relevance include the following:

- *Network density*: This metric represents the level of interconnectivity of the network in terms of relations between nodes. Density is expressed as a value between 0 and 1, with 0 representing a network with no connections between nodes and 1 representing a network in which every node is connected to every other concept (Kakimoto et al. 2006).
- *Sociometric status*: This metric provides a measure of how 'busy' a node is relative to the total number of nodes within the network under analysis (Houghton et al. 2006). Nodes with sociometric status values greater than the mean sociometric status value plus one standard deviation may be designated as 'key' (i.e. most connected) nodes.

- *Centrality*: This metric is a metric of the standing of a node within a network in terms of its distance from other nodes in the network (Houghton et al. 2006). A central node is one that is relatively close to all other nodes in the network in terms of connections; that is, an interaction with other nodes in the network is achieved through the smallest number of connections.

A.4.2 Hierarchical Task Analysis

A.4.2.1 Step 1: Define the Task under Analysis
The first step in applying HTA (Annett and Stanton 1998, Kirwan and Ainsworth 1992, Shepherd 1989) is to clearly define the task(s) under analysis and the boundaries surrounding the analysis.

A.4.2.2 Step 2: Data Collection
Once the task under analysis is clearly defined, specific data regarding the task should be collected to inform the development of the HTA. Data should be collected regarding the task steps involved, the technology used, interaction between humans, technology and team members. There are a number of ways to collect this data, including observations, concurrent verbal protocols, structured or semi-structured interviews (e.g. CDM) and questionnaires. The data collection approaches selected are dependent upon the various constraints imposed, such as time and access constraints.

A.4.2.3 Step 3: Determine the Overall Goal of the Task
The overall task goal of the task under analysis should first be specified at the top of the hierarchy (e.g. 'Land Boeing 737 at New Orleans Airport using the auto-land system' or 'make a cup of tea').

A.4.2.4 Step 4: Determine Task Sub-Goals
The next step of the HTA is to break the overall goal down into four or five meaningful sub-goals, which together make up the overall goal. For example, in an HTA analysis of bus driving (Salmon et al. 2011), the overall goal of 'operate bus' was broken down into the following sub-goals:

- Make pre-departure checks and adjustments.
- Drive bus.
- Deal with passengers.
- Maintain communications with TOC.
- Personal entertainment/communications/comfort tasks.

A.4.2.5 Step 5: Sub-Goal Decomposition
The sub-goals identified in step 4 should then be broken down into further sub-goals and operations, according to the task. This process continues until an appropriate sub-goal is reached. The bottom level of any branch in an HTA should always be an operation. Whereas everything above an operation specifies goals, operations actually state what needs to be done. Thus, operations are the actions to be made by the operator (whether human or technology).

A.4.2.6   Step 6: Plans Analysis

Once all of the sub-goals have been fully described, the plans need to be determined. Plans describe the order in which sub-goals and operations are undertaken to achieve the goals. A simple plan would say 'Do 1, then 2, and then 3'. Once the plan is completed, the operator returns to the super-ordinate level. Plans do not need to be linear and can come in any form such as 'Do 1, or 2 and 3'. Once the goals, sub-goals, operations and plans are exhausted, a complete diagram made up of these four aspects of the task makes up an HTA. If required, this can also be tabulated.

Further information about HTA can be found in the work of Stanton et al. (2013).

A.4.3   SYSTEMATIC HUMAN ERROR REDUCTION AND PREDICTION APPROACH

A.4.3.1   Step 1: Conduct an HTA

The first step in applying the Systematic Human Error Reduction and Prediction Approach (SHERPA; Embrey 1986) involves analysing the activity using HTA. As described in Section A.4.2, HTA is based upon the notion that task performance can be expressed in terms of a hierarchy of goals (what the operator or system is seeking to achieve), operations (the activities executed to achieve the goals) and plans (the sequence in which the operations are executed). The hierarchical structure of the analysis enables the analyst to progressively re-describe the activity in greater degrees of detail.

A.4.3.2   Step 2: Task Classification

Each operation from the bottom level of the HTA is taken in turn and is classified, using the SHERPA behaviour taxonomy, into one of the following:

- Action (e.g. pressing a button, engaging a piece of equipment, opening a door).
- Retrieval (e.g. retrieving information from a display or manual).
- Checking (e.g. conducting a check for signage).
- Selection (e.g. choosing one alternative over another).
- Information communication (e.g. exchanging information through verbal or non-verbal means).

A.4.3.3   Step 3: Human Error Identification

Once the task is classified, the analyst considers credible error modes associated with that activity, using the error taxonomy provided in Table A.1. For example, for tasks classified as actions, the action error modes A1–A10 are considered. Here the analyst asks whether each of the error modes could conceivably occur during the task in question. For each credible error, a description of the form that the error would take is provided (e.g. driver fails to see rail level crossing flashing lights).

A.4.3.4   Step 4: Consequence Analysis

Considering the consequence of each error on a system is an essential next step as the consequence has implications for the criticality of the error. The analyst should describe fully the consequences associated with the identified error. For example, the

TABLE A.1
SHERPA Error Taxonomy (Embrey, 1986)

Activity Type	Description
Action	• Operation too long/too short (A1)
	• Operation mistimed (A2)
	• Operation in wrong direction (A3)
	• Operation too little/too much (A4)
	• Misalign (A5)
	• Right operation on wrong object (A6)
	• Wrong operation on right object (A7)
	• Operation omitted (A8)
	• Operation incomplete (A9)
	• Wrong operation on wrong object (A10)
Checking	• Check omitted (C1)
	• Check incomplete (C2)
	• Right check on wrong object (C3)
	• Wrong check on right object (C4)
	• Check mistimed (C5)
	• Wrong check on wrong object (C6)
Retrieval	• Information not obtained (R1)
	• Wrong information obtained (R2)
	• Information retrieval incomplete (R3)
Communication	• Information not communicated (I1)
	• Wrong information communicated (I2)
	• Information communication incomplete (I3)
Selection	• Selection omitted (S1)
	• Wrong selection made (S2)

consequence of a driver not seeing the rail level crossing flashing lights is that the driver may not understand that a train is approaching the rail level crossing.

A.4.3.5 Step 5: Recovery Analysis

Next, the analyst should determine the recovery potential of the identified error. If the person making the error will realise immediately, then 'immediate' is entered. If the recovery is not immediate, but there is a later task step at which the error could be recovered, the task step number from the HTA is entered. If there is no recovery step, then 'None' is entered.

A.4.3.6 Step 6: Ordinal Probability Analysis

Once the consequence and recovery potential have been identified, the analyst rates the probability of the error occurring. An ordinal probability value is entered as low, medium or high. If the error has never been known to occur, then a low (L) probability is assigned. If the error has occurred on previous occasions, the medium (M) probability is assigned. Finally, if the error occurs frequently, a high (H) probability is assigned. This analysis often relies upon historical data and/or input from SMEs.

A.4.3.7   Step 7: Criticality Analysis

Criticality is assigned in a binary manner. If the error would lead to a serious incident, then it is classified as critical. Typically, a critical consequence would be one that would lead to substantial human costs (e.g. injuries, fatalities), economic costs (e.g. property damage, loss of operating capacity) or environmental damage (e.g. irreversible environmental impacts).

A.4.3.8   Step 8: Identification of Remedial Measures

The final stage in the process is to propose error reduction strategies. These are presented in the form of suggested changes to the work system that could have prevented the error from occurring or reduced the consequences of the error. This typically occurs through a structured brainstorming exercise where the analyst(s) propose ways of improving system design through: reducing the likelihood of the error occurring, enabling operators to identify when they have made an error, enabling recovery from error, or mitigating the consequences of the error. Such strategies might include changes to equipment, training, standard operating procedures, organisational policy or culture.

Some remedies may be very costly to implement. Therefore, they need to be assessed with regard to the consequences, criticality and probability of the error.

Further information about using SHERPA can be found in the work of Stanton et al. (2013).

A.4.4   Cognitive Work Analysis

A.4.4.1   Step 1: Determine the Boundaries of the Analysis

The first step in a CWA is to clearly define the aims of the analysis and the boundaries of the system under consideration to allow for an appropriate description of the system.

A.4.4.2   Step 2: Select Appropriate CWA Phases and Methods

Once the nature and desired outputs of the analysis are clearly defined, the analysis team carefully selects the most appropriate CWA phases and methods to be employed during the analysis. For example, when using the framework for the design of a novel interface, it may be that only the Work Domain Analysis (WDA) component is required. In general, it is recommended that WDA be applied as a starting point as it provides a holistic view of the system. Based on the selection of phases and methods, steps 3–8 are conducted as appropriate.

It should be noted that it is beyond the scope of this book to fully describe the CWA procedure. The following guidance is intended to act as a broad set of guidelines for each of the phases defined by the CWA framework. A more complete description can be found in the work of Jenkins et al. (2009) or Vicente (1999).

A.4.4.3   Step 3: Conduct Work Domain Analysis

The initial phase within the CWA framework, WDA, provides a description of the constraints that govern the purpose and the function of the systems under analysis. The abstraction hierarchy (Rasmussen 1986, Vicente 1999) is used to provide a description of the domain that is independent of context and independent of actors. The analysis, and the resultant set of diagrams, is not specific to any particular technology; rather, they represent the entire domain. The top three levels of the abstraction hierarchy identify the overall objectives of the domain, and what it can achieve, whereas the bottom two levels concentrate on the physical components and their affordances. Through a series of functional 'means-ends' links, it is possible to analyse how individual components can have an impact on the overall domain purpose. The abstraction hierarchy is constructed by considering the work system's objectives (top-down) and the work system's capabilities (bottom-up). The diagram is constructed based upon a range of data collection opportunities. The exact data collection procedure is dependent on the domain in question and the availability of data. In most cases, document analysis is used as a starting point. Document analysis allows the analyst to gain a basic domain understanding, forming the basis for semi-structured interviews with domain experts. Wherever possible, observation of the activity in context is highly recommended.

The abstraction hierarchy consists of five levels of abstraction, ranging from the most abstract level of purposes to the most concrete level of form (Vicente 1999). The labels used for each of the levels of the hierarchy tend to differ, depending on the aims of the analysis. The following are the generic labels for the hierarchical levels.

- *Functional purpose*: The domain purpose(s), displayed at the very top of the diagram, represents the reason why the work system exists. This purpose is independent of any specific situation and is also independent of time; the system purpose exists as long as the system does.
- *Values and priority measures*: This level captures the key values that can be used to assess how well the work system is achieving its domain purpose(s).
- *Generalised functions*: The middle layer of the hierarchy identifies the functions that are performed within the system for it to meet its purposes.
- *Physical processes*: This level identifies the physical process or 'affordances' that each physical object provides. These are listed generically and are independent of the domain purpose.
- *Physical objects*: The physical objects within the system are identified at the base of the hierarchy. These objects represent the relevant objects from all of the component technologies. This level of the diagram is independent of purpose; however, analyst judgement is required to limit the object list to a manageable size.

The structure of the abstraction hierarchy framework acts as a guide to acquiring the knowledge necessary to understand the domain. The framework helps to direct the search for deep knowledge about the system, providing structure to the document analysis process, particularly for the domain novice. Although the output may initially appear overbearing, its value to the analysis cannot be overstated. The abstraction hierarchy defines the systemic constraints at the highest level.

The WDA can also be described using an abstraction decomposition space (ADS). This is represented as a matrix which represents the levels of abstraction on the vertical dimension and the levels of decomposition across the horizontal dimension. The decomposition hierarchy typically decomposes the system according to five levels of resolution, ranging from the coarsest level of the overall system to the finest level of component (Vicente 1999). According to Vicente (1999), each of the five levels represents a different level of granularity with respect to the system in question. Moving from left to right across the decomposition hierarchy is the equivalent of zooming into the system, as each level across represents a more detailed representation. The ADS also employs structural means-ends relationships in order to link the different elements of the system within the ADS. This means that every node in the ADS should be the end that is achieved by all of the nodes below it, and also the means that can be used to achieve all of the nodes above it.

A.4.4.4   Step 4: Conduct Control Task Analysis – Contextual Activity Template

To this point, the analysis does not deliberately consider the constraints that are imposed by specific situations. A tool for considering such constraints is the contextual activity template (Naikar et al. 2006). This tool plots the functions identified in the abstraction hierarchy against a number of specific 'situations'. At this stage, the analysis remains independent of the actor. That is, the focus is on functions specified generally, not on tasks performed by any individual actor (i.e. person or technological actor). The first stage of the process is to define the situations. Situations can be characterised by either time or location, or a combination of the two. In many cases, it is appropriate to explore more than one set of situations using multiple contextual activity template representations to meet a range of analytic goals.

Generally, the contextual activity template takes the generalised functions from the abstraction hierarchy and adds information by describing the situational constraints on the achievement of these functions. Thus, the products provide a more context-specific description of the domain.

A.4.4.5   Step 5: Conduct Control Task Analysis – Decision Ladders

Continuing with the theme of describing additional constraints, key function–situation cells within the contextual activity template can be explored in terms of decision-making. The decision ladder (Rasmussen et al. 1994) is the tool most commonly used within CWA to describe the decision-making activity. Its focus is on the entire decision-making activity rather than the moment of selection between options. It is not specific to any single actor; instead, it represents the decision-making process of the combined work system. In many cases, the decision-making process may be collaborative, distributed between a range of human and technical decision-makers.

The ladder contains two different types of nodes: rectangular boxes that represent data-processing activities and circles that represent resultant states of knowledge. The left side of the decision ladder represents the observation of the current system state, whereas the right side represents the planning and execution of tasks and procedures to achieve a target system state. Novice users (to the situation) are expected to follow the decision ladder in a linear fashion, whereas expert users are expected to take 'shortcuts' through the ladder.

Decision ladders can be populated based on CDM interviews or other semi-structured interviews with SMEs. Decision ladder models can be used to identify the information requirements for making a decision triggered by a number of presupposed events. However, at this stage of the analysis, the relative importance of these information elements is not considered.

A.4.4.6 Step 6: Conduct Strategies Analysis

Strategies Analysis addresses the constraints influencing the way in which an activity can be conducted. In keeping with the remainder of the framework, it introduces an additional level of detail to the analyses in the previous phases. The aim of a Strategies Analysis is to describe the constraints that dictate how a system can be (rather than how it should be or currently is) moved from one state to another. This phase of the analysis can be particularly useful for exploring flexibility within a system (Jenkins et al. 2009). Although information flow maps are typically used for the Strategies Analysis component of CWA, other tools, such as the strategies analysis diagram (Cornelissen et al. 2013), can also be applied. The strategies analysis diagram builds on the abstraction hierarchy developed in the WDA phase and involves the addition of two levels to the diagram: verbs and criteria. The verbs are used to specify how the physical objects can be used. The criteria are then used to specify the circumstances under which different strategies might be chosen.

A.4.4.7 Step 7: Conduct Social Organisation and Cooperation Analysis

The Social Organisation and Cooperation Analysis phase of CWA involves identifying how activities and strategies can be distributed between actors (human and technological) within the system. This phase of CWA reuses the outputs such as the ADS, decision ladders and information flow maps developed during the preceding phases. It involves mapping onto these previous phases which actors are currently involved in contributing to functions, decisions or strategies, and it can be used to explore how task allocation could be changed.

A.4.4.8 Step 8: Conduct Worker Competencies Analysis

The final stage of a CWA involves identifying the cognitive skills required for task performance. Worker Competencies Analysis uses Rasmussen's (1983) skills, rules and knowledge framework to classify the cognitive activities employed by agents during control task performance. The skills, rules and knowledge taxonomy (Rasmussen 1983) enables the analyst to map, for each critical strategy, how the system supports each level of information processing.

In the skills, rules and knowledge model, skill-based behaviour is associated with sensory motor performance that occurs in skilled activity without conscious control being required. Rule-based behaviour refers to the application of stored rules, based on the past experience, to determine behaviour. Finally, knowledge-based behaviour is engaged in unfamiliar situations where it is not possible to draw upon the past experience and the actor must engage in reasoning to understand the situation and select an appropriate course of action.

Further guidance for using the CWA framework can be found in the works of Vicente (1999), Naikar (2013) and Jenkins et al. (2009).

A.5   HUMAN FACTORS DESIGN METHODS

This section provides guidance for using a selection of key human factors design methods. These methods can be used following the application of systems-based analysis methods, such as CWA, as recommended by the CWA Design Toolkit (CWA-DT) process (see Chapter 6 for more information about the CWA-DT).

A.5.1   Scenarios and Stories

A.5.1.1   Step 1: Identify the Key Themes for Communication

Following the analysis process, key themes or insights should be identified that are important to communicate to design participants (e.g. users or stakeholders participating in a design process or other members of a design team). These insights can be communicated using use scenarios (fictional descriptions of user interaction with a system) or stories (documented real-life interactions with the system). Key themes may relate to particular issues faced by participants (described in CWA-DT as 'pain points'), or they could reflect a series of typical system encounters. These design tools can also provide a way for design participants to gain empathy with system users.

A.5.1.2   Step 2: Develop Scenarios and/or Stories

Once the themes have been identified, the narratives should be developed. They may range from short descriptions (e.g. one paragraph) to longer (e.g. one page). The length will depend on the level of detail needed to communicate the key points and issues, and to develop a sense of empathy with the user(s) involved in the scenario.

The content of scenarios can draw directly from the CWA outputs. For example, the user goals represented in scenarios could be drawn from decision ladders, the artefacts from the WDA, and contextual information from the situations in the contextual activity template. Further, if using the CWA-DT, insights documented regarding scenario features should be incorporated where possible to ensure a strong link between the analysis and the scenarios.

To promote the development of empathy with actors, it is recommended that scenarios be only partially specified. For example, the scenario may describe the situation of a user, including their personal attributes, the goal they are working towards and how they interact with objects in the system. However, there may be gaps left in the narrative prompting the participant to consider and describe how the user feels at different points in the scenario (e.g. frustrated, pressured, delighted) and what they have learned about the system through their interaction.

Stories, on the other hand, can be derived directly from data collection activities. For example, stories may be used to convey the details of an interaction gathered during a CDM interview or other SME interaction. If using stories, it is important to ensure that the case is de-identified to avoid breaching participant confidentiality.

As an alternative to written narrative form, scenarios and stories could be documented in video form, or as storyboards. If using a narrative, consider whether photographs or other images could be used to enhance the realism of the description.

A.5.1.3   Step 3: Present Scenarios/Stories to Design Participants

If using scenarios or stories in a workshop context, the following process is recommended. Firstly, the scenario/story should be briefly introduced to participants, and they should be given time to review the narrative. If the narratives are not fully specified, participants are also asked to complete the gaps.

Then, participants can be asked to discuss the following types of topics in small groups:

- What values were represented in the scenario?
- What goals were represented in the scenario?
- What issues or problems were apparent in the scenario?
- What opportunities were apparent in the scenario?

The process is repeated with subsequent narratives as required, based on the number of key themes to be communicated.

A.5.1.4   Step 4: Document Design Ideas and Insights

At the conclusion of the exercise, groups should be asked to share design insights that emerged from their discussions. These should be documented and may be used in further design activities.

A.5.1.5   Step 5: Evaluate and Refine Design Ideas

Scenarios and stories can also be used to evaluate early design concepts to determine how changes to a system might affect the user experience. This can identify potential issues or refinements needed to ensure that changes do not introduce unintended negative effects for users.

Further information about the use of scenarios and stories in design can be found in the works of Carroll (2002) and Erickson (1995), respectively.

A.5.2   Personas

A.5.2.1   Step 1: Identify Characters

Like scenarios and stories, personas can help design participants to gain a better understanding of system users and develop empathy for different types of users. The first step in using personas is to identify which 'characters' will be used. It is advisable to use some personas that represent typical system users and some that represent more extreme users (as suggested by Djajadiningrat et al. 2000). In addition, characters may not only represent system users but also other stakeholders such as maintainers or managers. This assists to create designs that meet the needs of both users and businesses, and ensures that considerations across the life cycle are incorporated into designs.

A.5.2.2   Step 2: Develop Personas

Once the type of characters has been determined, the personas can be developed. These can be provided in written form or by other means such as video. The personas should provide personal information about the character and information relevant to their interactions with the system. For example, personas about drivers might incorporate the following:

- Demographic characteristics such as age, gender and years of driving experience.
- Main reason for travel such as commuting and recreation.
- Personal details such as employment type, area of residence (urban, rural) and family commitments.

Personas might then provide information about the character's perceptions of driving, such as what concerns or challenges them about driving or what they enjoy about driving. An alternative method is to pose these as questions, for design participants to complete based on the general information provided. Asking participants to complete this information requires them to put themselves into the character's shoes. Alternatively, design participants might be asked to perform a role play as a character who is undertaking a task relating to the design process. This provides another means for participants to develop empathy with users and system stakeholders.

A.5.2.3   Step 3: Present Personas to Design Participants

If using personas in a workshop context, the following process is recommended. Firstly, provide each persona to participants and enable them to review the narrative. If questions are posed, provide time for participants to complete these individually. Then participants should share their answers in small groups and discuss how their perceptions of the characters might influence design.

Repeat with subsequent personas as required, based on the number of key themes to be communicated.

A.5.2.4   Step 4: Document Design Ideas and Insights

At the conclusion of the exercise, groups should be asked to share design insights that emerged from their discussions. These should be documented and may be used in further design activities.

A.5.3   Inspiration Cards

A.5.3.1   Step 1: Identify the Insights or Themes to Use with Cards

The first step in using inspiration cards in design is to assess the insights gained from conducting systems-based analysis and identify leverage points, pain points and other types of design opportunities. Select six to eight of these insights based on their importance to address, or to explore, for improving system design.

A.5.3.2   Step 2: Prepare Inspiration Cards

Once the leverage points, pain points and other design opportunities have been identified, cards should be created to represent the selected insights. Cards should be approximately palm sized to enable easy handling and sorting.

Next, sets of design inspiration cards should be reviewed and a range of cards relevant to the design question should be selected for use. Design toolkits could include the Design with Intent Toolkit (Lockton et al. 2010), new technology cards (Halskov and Dalsgard 2006) or sign cards (Brandt and Messerter 2004).

In the selection process, it is advisable to choose cards that align with the overall design philosophy underpinning the research. For example, the Design with Intent Toolkit includes design patterns under the 'Machiavellian lens', which focusses on meeting the needs of system designers regardless of the needs of users. This lens includes ideas such as functional obsolescence, whereby the card asks 'Can you design things to become technologically superseded (or even wear out) quickly, so people replace them?' These types of design interventions may be contrary to sociotechnical systems theory philosophies and so may not be appropriate for design processes applying this philosophy.

A.5.3.3 Step 3: Use of Cards

In a workshop setting, provide the insight cards and the design inspiration cards to participants working in small groups. Ask participants to work by sorting through the cards and exploring whether any of the insight cards can be matched with a design card and, if so, to brainstorm potential design ideas based on the combination.

A.5.3.4 Step 4: Document Design Ideas and Insights

At the conclusion of the exercise, groups should be asked to share design insights that emerged from their discussions. These should be documented and may be used in further design activities.

A.5.4 Assumption Crushing

A.5.4.1 Step 1: Identify Assumptions

Assumptions include theories, beliefs or hypotheses underpinning the structure of the system and the way things are currently done. They may not be consciously realised but can unconsciously restrict the type of design ideas that are considered. Crushing assumptions means to identify an alternative or opposite theory, belief or hypothesis and brainstorm design ideas in line with this (Imber 2012). Assumptions uncovered during the analysis process should be reviewed, or the CWA-DT prompt questions (Read et al. 2016) can be used to review the analyses and identify assumptions. Once assumptions are identified, two to three should be selected. Those selected should generally be those that are most core to the underlying design of the current system.

A.5.4.2 Step 2: Crush Assumptions

Once the set of assumptions to be crushed has been identified, these are presented to participants for consideration. Participants are asked to 'crush' the assumption by individually brainstorming an alternative statement for the assumption. These alternative statements should be shared with the wider group, and then one alternative statement should be selected for each assumption and participants asked to work in small groups and brainstorm design ideas in line with that alternative statement.

A.5.4.3 Step 3: Document Design Ideas and Insights

At the conclusion of the exercise, groups should be asked to share design insights that emerged from their discussions. These should be documented and may be used in further design activities.

A.5.5   Metaphors and Analogies

A.5.5.1   Step 1: Identify Metaphors and/or Analogies

Metaphors and analogies provide a means of seeing things in a new way that can prompt new ideas and innovation. The use of metaphors can also assist the translation of conventions from one area to another, providing end users with a familiar model to understand a new system. Metaphors or analogies uncovered during the analysis process should be reviewed, or the CWA-DT prompt questions (Read et al. 2016) can be used to review the analyses and identify potential metaphors for use in prompting design ideas. Where a number of metaphors have been uncovered, it is advisable to select a small number (e.g. two to three) based on those the analyst thinks will be most fruitful in generating novel ideas (i.e. have a strong connection to the initial system, yet are different enough to provoke new connections between the two subjects).

A.5.5.2   Step 2: Use of Metaphors

The selected metaphors and/or analogies are presented to participants who are asked to discuss these and then to consider how the existing system could be changed to be similar to the metaphor or analogy. One type of metaphor involves viewing the domain of interest from the perspective of a different domain. For example, in work on public transport ticketing systems (Read et al. 2015a), we identified the analogy of airline ticketing and asked how features in that domain could be transferred across. For example, frequent flyer points and online check-in options were considered to determine if they might be useful in the public transport domain.

It is advisable to have participants consider the metaphors individually first, then to work in small groups to share their ideas and build upon them.

A.5.5.3   Step 3: Document Design Ideas and Insights

At the conclusion of the exercise, groups should be asked to share design insights that emerged from their discussions. These should be documented and may be used in further design activities.

Further information about the use of metaphors and analogies in design can be found in the work of Madsen (1994).

References

Ablon, L., M. C. Libicki, and A. A. Golay. 2014. *Markets for Cybercrime Tools and Stolen Data*. Santa Monica, CA: Rand Corporation.

Ahlstrom, U. 2005. Work domain analysis for air traffic controller weather displays. *Journal of Safety Research* 36 (2): 159–169.

Ainsworth, L., and E. Marshall. 1998. Issues of quality and practicality in task analysis: Preliminary results from two surveys. *Ergonomics* 41 (11): 1604–1617. Reprinted in *Task Analysis*, edited by J. Annett and N. A. Stanton. 2000, pp. 79–89. London: Taylor & Francis.

Annett, J. 2004. Hierarchical task analysis. In *The Handbook of Task Analysis for Human-Computer Interaction*, edited by J. Diaper and N. A. Stanton, pp. 67–82. Mahwah, NJ: Lawrence Erlbaum Associates.

Annett, J., D. J. Cunningham, and P. Mathias-Jones. 2000. A method for measuring team skills. *Ergonomics* 43 (8): 1076–1094.

Annett, J., and N. A. Stanton. 1998. Research and developments in task analysis. *Ergonomics* 41 (11): 1529–1536. Reprinted in *Task Analysis*, edited by J. Annett and N. A. Stanton. 2000, pp. 1–8. London: Taylor & Francis.

Appelbaum, S. H. 1997. Socio-technical systems theory: An intervention strategy for organizational development. *Management Decision* 35 (6): 452–463.

Baber, C., and N. A. Stanton. 1994. Task analysis for error identification: A methodology for designing error-tolerant consumer products. *Ergonomics* 37 (11): 1923–1941.

Badham, R. J., C. W. Clegg, and T. Wall. 2006. Sociotechnical theory. In *International Encyclopedia of Ergonomics and Human Factors*, 2nd ed., edited by W. Karwowski, pp. 2347–2350. Boca Raton, FL: CRC Press.

Banks, V. A., and N. A. Stanton. 2016. Keep the driver in control: Automating automobiles of the future. *Applied Ergonomics* 53, Part B: 389–395.

Banks, V. A., N. A. Stanton, and C. Harvey. 2014. Sub-systems on the road to vehicle automation: Hands and feet free but not 'mind' free driving. *Safety Science* 62: 505–514.

Beanland, V., and E. H. Chan. 2016. The relationship between sustained inattentional blindness and working memory capacity. *Attention, Perception, & Psychophysics* 76 (3): 808–817.

Beanland, V., M. G. Lenné, P. M. Salmon, and N. A. Stanton. 2013. A self-report study of factors influencing decision-making at rail level crossings: Comparing car drivers, motorcyclists, cyclists and pedestrians. In *Proceedings of the 2013 Australasian Road Safety Research, Policing & Education Conference*, pp. 1–11. Canberra, Australia: Australasian College of Road Safety.

Beanland, V., M. G. Lenné, P. M. Salmon, and N. A. Stanton. 2016. Variability in decision-making and critical cue use by different road users at rail level crossings. *Ergonomics* 59 (6): 754–766.

Bisantz, A. M., and C. M. Burns. 2008. *Applications of Cognitive Work Analysis*. Boca Raton, FL: CRC Press.

BITRE (Bureau of Infrastructure, Transport and Regional Economics). 2016. *Road Deaths Australia, November 2016*. Canberra, Australia: Commonwealth of Australia.

Brandt, E., and J. Messerter. 2004. Facilitating collaboration through design games. *Proceedings of the Eighth Conference on Participatory design: Artful Integration: Interweaving Media, Materials and Practices* 1: 121–131.

Bredemeier, K., and D. J. Simons. 2012. Working memory and inattentional blindness. *Psychonomic Bulletin & Review* 19 (2): 239–244.

Brooke, J. 1996. SUS: A 'quick and dirty' usability scale. In *Usability Evaluation in Industry*, edited by P. W. Jordan, B. Thomas, B. A. Weerdmeester, and I. L. McClelland, pp. 189–194. London: Taylor & Francis.

Burdett, B. R. D., S. G. Charlton, and N. J. Starkey. 2016. Not all minds wander equally: The influence of traits, states and road environment factors on self-reported mind wandering during everyday driving. *Accident Analysis & Prevention* 95, Part A: 1–7.

Burns, C. M., D. J. Bryant, and B. A. Chalmers. 2005. Boundary, purpose, and values in work domain models: Models of naval command and control. *IEEE Transactions on Systems, Man, and Cybernetics* 35: 603–616.

Burns, C. M., and J. R. Hajdukiewicz. 2004. *Ecological Interface Design*. Boca Raton, FL: CRC Press.

Burns, C. M., G. Skraaning, G. A. Jamieson, N. Lau, J. Kwok, R. Welch, and G. Andresen. 2008. Evaluation of ecological interface design for nuclear process control: Situation awareness effects. *Human Factors* 50: 663–679.

Cairney, P., T. Gunatillake, and E. Wigglesworth. 2002. *Reducing Collisions at Passive Railway Level Crossings in Australia*. Sydney, Australia: Austroads.

Cale, M. H., A. Gellert, N. Katz, and W. Sommer. 2013. Can minor changes in the environment lower accident risk at level crossings? Results from a driving simulator-based paradigm. *Journal of Transportation Safety & Security* 5 (4): 344–360.

Carroll, J. M. 2002. Scenario-based design. In *International Encyclopedia of Ergonomics and Human Factors*, 2nd ed., edited by W. Karwowski. Boca Raton, FL: CRC Press.

Chapman, P. R., and G. Underwood. 1998. Visual search of dynamic scenes: Event types and the role of experience in viewing driving situations. In *Eye Guidance in Reading and Scene Perception*, edited by G. Underwood, pp. 369–392. Oxford: Elsevier.

Checkland, P. 1981. *Systems Thinking, Systems Practice*. Chichester: John Wiley & Sons.

Cherns, A. 1976. The principles of sociotechnical design. *Human Relations* 29 (8): 783–792.

Cherns, A. 1987. Principles of sociotechnical design revisited. *Human Relations* 40 (3): 153–161.

Clacy, A., N. Goode, R. Sharman, G. P. Lovell, and P. M. Salmon. 2016. A knock to the system: A new sociotechnical systems approach to sport-related concussion. *Journal of Sports Sciences* 1–8.

Clark, H. E., J. A. Perrone, R. B. Isler, and S. G. Charlton. 2016. The role of eye movements in the size-speed illusion of approaching trains. *Accident Analysis & Prevention* 86: 146–154.

Clegg, C. W. 2000. Sociotechnical principles for system design. *Applied Ergonomics* 31 (5): 463–477.

Conti, J., T. B. Sheridan, and J. Multer. 1998. Experimental evaluation of retroreflective markings on rail cars at highway-railroad grade crossings. *Paper presented at the 5th International Symposium on Railroad-Highway Grade Crossing Research and Safety*, Knoxville, TN, October.

Cornelissen, M., P. M. Salmon, D. P. Jenkins, and M. G. Lenné. 2013. A structured approach to the strategies analysis phase of cognitive work analysis. *Theoretical Issues in Ergonomics Science* 14 (6): 546–564.

Cornelissen, M., P. M. Salmon, N. A. Stanton, and R. McClure. 2015. Assessing the 'system' in safe systems-based road designs: Using cognitive work analysis to evaluate intersection designs. *Accident Analysis and Prevention* 74: 324–338.

Cornelissen, M., P. M. Salmon, and K. L. Young. 2012. Same but different? Understanding road user behaviour at intersections using cognitive work analysis. *Theoretical Issues in Ergonomics Science* 14 (6): 592–615.

Coroner Hendtlass. 2013. *Coronial Investigation of Twenty-Six Rail Crossing Deaths in Victoria, Australia*. Melbourne, Australia: Coroners Court of Victoria.

Crandall, B., G. Klein, and R. R. Hoffman. 2006. *Working Minds: A Practitioner's Guide to Cognitive Task Analysis*. Cambridge, MA: MIT Press.

Curry, A. E., J. Hafetz, M. J. Kallan, F. K. Winston, and D. R. Durbin. 2011. Prevalence of teen driver errors leading to serious motor vehicle crashes. *Accident Analysis & Prevention* 43 (4): 1285–1290.

Davey, J. D., N. R. Ibrahim, and A. M. Wallace. 2005. Motorist behaviour at railway level crossings: An exploratory study of train driver experience. *Paper presented at the Australasian Road Safety Research, Policing and Education Conference*, Wellington, New Zealand, November 14–16.

Davis, F. D. 1989. Perceived usefulness, perceived ease of use, and user acceptance of information technology. *MIS Quarterly* 13 (3): 319–340.

Davis, L. E. 1982. Organization design. In *Handbook of Industrial Engineering*, edited by G. Salvendy, pp. 2.1.1–2.1.29. New York: John Wiley & Sons.

Davis, M. C., R. Challenger, D. N. W. Jayewardene, and C. W. Clegg. 2014. Advancing sociotechnical systems thinking: A call for bravery. *Applied Ergonomics* 45 (2), Part A: 171–180.

Davis, M. C., D. J. Leach, and C. W. Clegg. 2011. The physical environment of the office: Contemporary and emerging issues. In *International Review of Industrial and Organizational Psychology 2011*, edited by G. P. Hodgkinson and J. K. Ford, pp. 193–237. New York: John Wiley & Sons.

Dekker, S. 2011. *Drift into Failure: From Hunting Broken Components to Understanding Complex Systems*. Farnham: Ashgate.

Department for Transport. 2016. *Reported Road Casualties in Great Britain: Quarterly Provisional Estimates Year Ending June 2016*. London: Department for Transport.

Djajadiningrat, J. P., W. W. Gaver, and J. W. Frens. 2000. Interaction relabelling and extreme characters: Methods for exploring aesthetic interactions. In *Proceedings of the 3rd Conference on Designing Interactive Systems: Processes, Practices, Methods, and Techniques*, edited by D. Boyarski and W. A. Kellogg, pp. 66–71. New York: ACM.

Dul, J., R. Bruder, P. Buckle, P. Carayon, P. Falzon, W. S. Marras, J. R. Wilson, and B. van der Doelen. 2012. A strategy for human factors/ergonomics: Developing the discipline and profession. *Ergonomics* 55 (4): 377–395.

Eason, K. D. 1991. Ergonomic perspectives in advances in human-computer interaction. *Ergonomics* 34 (6): 721–741.

Eason, K. 2008. Sociotechnical systems theory in the 21st century: Another halffilled glass? In *Sense in Social Science: A Collection of Essays in Honour of Dr. Lisl Klein 2008*, edited by D. Graves, pp. 123–134. Broughton: Desmond Grave.

Eason, K. 2014. Afterword: The past, present and future of sociotechnical systems theory. *Applied Ergonomics* 45 (2): 213–220.

Edquist, J., K. Stephan, E. Wigglesworth, and M. G. Lenné. 2009. *A Literature Review of Human Factors Safety Issues at Australian Level Crossings*. Melbourne, Australia: Monash University Accident Research Centre.

Embrey, D. E. 1986. SHERPA: A systematic human error reduction and prediction approach. *Paper presented at the International Meeting on Advances in Nuclear Power Systems*, Knoxville, TN, April 21–24.

Erickson, T. 1995. Notes on design practice: Stories and prototypes as catalysts for communication. In *Scenario-based Design: Envisioning Work and Technology in System Development*, edited by J. M. Carroll, pp. 37–58. New York: John Wiley & Sons.

Falkmer, T., and N. P. Gregersen. 2005. A comparison of eye movement behavior of inexperienced and experienced drivers in real traffic environments. *Optometry & Vision Science* 82 (8): 732–739.

Flach, J. M., R. G. Eggleston, G. G. Kuperman, and C. O. Dominguez. 1998. *SEAD and the UCAV: A Preliminary Cognitive Systems Analysis (AFRL-HE-WP-TR-1998-0013)*. Wright-Patterson Air Force Base, OH: AFRL, Human Effectiveness Directorate.

Flanagan, J. C. 1954. The critical incident technique. *Psychological Bulletin* 51 (4): 327–358.

Galanter, C. A., and V. L. Patel. 2005. Medical decision making: A selective review for child psychiatrists and psychologists. *Journal of Child Psychology and Psychiatry* 46 (7): 675–689.

Gershon, P., and D. Shinar. 2013. Increasing motorcycles attention and search conspicuity by using Alternating-Blinking Lights System (ABLS). *Accident Analysis & Prevention* 50: 801–810.

Gibson, J. J., and L. E. Crooks. 1938. A theoretical field-analysis of automobile-driving. *The American Journal of Psychology* 51 (3): 453–471.

Goh, Y. M., and P. E. D. Love. 2012. Methodological application of system dynamics for evaluating traffic safety policy. *Safety Science* 50 (7): 1594–1605.

Golightly, D., and N. Dadashi. 2017. The characteristics of railway service disruption: Implications for disruption management. *Ergonomics* 60 (3): 307–320.

Goode, N., G. J. M. Read, M. R. H. van Mulken, A. Clacy, and P. M. Salmon. 2016. Designing system reforms: Using a systems approach to translate incident analyses into prevention strategies. *Frontiers in Psychology* 7 (1974): 1–17.

Green, M. 2002. Signs & signals. *Occupational Health & Safety* 71 (6): 30–32, 34, 36.

Gregory, J. 2003. Scandanavian approaches to participatory design. *International Journal of Engineering Education* 19 (1): 62–74.

Gstalter, H., and W. Fastenmeier. 2010. Reliability of drivers in urban intersections. *Accident Analysis & Prevention* 42 (1): 225–234.

Halskov, K., and P. Dalsgård. 2006. Inspiration card workshops. In *Proceedings of the 6th Conference on Designing Interactive Systems*, edited by J. M. Carroll, S. Bødker, and J. Coughlin, pp. 2–11. New York: ACM.

Harris, D., N. A. Stanton, A. Marshall, M. S. Young, and P. Salmon. 2005. Using SHERPA to predict design-induced error on the flight deck. *Aerospace Science and Technology* 9 (6): 525–532.

Hart, S. G., and L. E. Staveland. 1988. Development of a Multi-dimensional workload rating scale: Results of empirical and theoretical research. In *Human Mental Workload*, edited by P. A. Hancock and N. Meshkati. Amsterdam, the Netherlands: Elsevier.

Hatakka, M., E. Keskinen, N. P. Gregersen, A. Glad, and K. Hernetkoski. 2002. From control of the vehicle to personal self-control: Broadening the perspectives to driver education. *Transportation Research Part F: Traffic Psychology and Behaviour* 5 (3): 201–215.

Heinrich, H. W. 1931. *Industrial Accident Prevention: A Scientific Approach*. New York: McGraw-Hill.

Hettinger, L. J., A. Kirlik, Y. M. Goh, and P. Buckle. 2015. Modelling and simulation of complex sociotechnical systems: Envisioning and analysing work environments. *Ergonomics* 58 (4): 600–614.

Highways Agency. 2012. *Design Manual for Roads and Bridges*. London: Highways Agency.

Hirschhorn, L., P. Noble, and T. Rankin. 2001. Sociotechnical systems in an age of mass customization. *Journal of Engineering and Technology Management* 18 (3–4): 241–252.

Hollnagel, E. 2012. *FRAM: The Functional Resonance Analysis Method: Modelling Complex Socio-technical Systems*. Farnham: Ashgate.

Holmqvist, K., M. Nyström, R. Andersson, R. Dewhurst, H. Jarodzka, and J. van de Weijer. 2011. *Eye Tracking: A Comprehensive Guide to Methods and Measures*. Oxford: Oxford University Press.

Houghton, R. J., C. Baber, R. McMaster, N. A. Stanton, P. M. Salmon, R. Stewart, and G. H. Walker. 2006. Command and control in emergency services operations: A social network analysis. *Ergonomics* 49 (12–13): 1204–1225.

Imber, A. 2009. *The Creativity Formula*. Caulfield, Australia: Liminal Press.

Imber, A. 2012. *Five Ways to Boost Creativity*. Retrieved from http://www.inventium.com.au/wp-content/uploads/2013/06/5-Ways-to-Boost-Creativity-v1.pdf (Accessed March 11, 2017).

ISO/TR 16982:2002. *Ergonomics of Human-System Interaction – Usability Methods Supporting Human-Centred Design*. Geneva, Switzerland: International Organization for Standardization.

ITSR (Independent Transport Safety Regulator). 2011. *Managing Signals Passed at Danger*. Sydney, Australia: Independent Transport Safety Regulator.

Jansson, A., E. Olsson, and M. Erlandsson. 2006. Bridging the gap between analysis and design: Improving existing driver interfaces with tools from the framework of cognitive work analysis. *Cognition, Technology & Work* 8 (1): 41–49.

Jenkins, D. P., P. M. Salmon, N. A. Stanton, and G. H. Walker. 2010. A new approach for designing cognitive artefacts to support disaster management. *Ergonomics* 53 (5): 617–635.

Jenkins, D. P., N. A. Stanton, P. M. Salmon, and G. H. Walker. 2009. *Cognitive Work Analysis: Coping with Complexity*. Farnham: Ashgate.

Kakimoto, T., Y. Kamei, M. Ohira, and K. Matsumoto. 2006. Social network analysis on communications for knowledge collaboration in OSS communities. In *Proceedings of the 2nd International Workshop on Supporting Knowledge Collaboration in Software Development (KCSD'06)*, edited by Y. Ye and M. Ohira, pp. 35–41. Tokyo, Japan: National Institute of Informatics.

Kee, D., G. T. Jun, P. Waterson, and R. Haslam. 2017. A systemic analysis of South Korea Sewol ferry accident – Striking a balance between learning and accountability. *Applied Ergonomics* 59, Part B: 504–516.

King, M. J., D. Soole, and A. Ghafourian. 2009. Illegal pedestrian crossing at signalised intersections: Incidence and relative risk. *Accident Analysis & Prevention* 41 (3): 485–490.

Kirwan, B., and L. Ainsworth. 1992. *A Guide to Task Analysis*. London: Taylor & Francis.

Klein, G. 1993. *Naturalistic Decision Making: Implications for Design, State-of-the-Art Report*. Alexandria, VA: Crew System Ergonomics Information Analysis Center.

Klein, G., and A. A. Armstrong. 2005. Critical decision method. In *Handbook of Human Factors and Ergonomics Methods*, edited by N. Stanton, A. Hedge, K. Brookhuis, E. Salas, and H. Hendrick, pp. 347–356. Boca Raton, FL: CRC Press.

Klein, G. A., R. Calderwood, and D. MacGregor. 1989. Critical decision method for eliciting knowledge. *IEEE Transactions on Systems, Man and Cybernetics* 19 (3): 462–472.

Kreitz, C., P. Furley, D. Memmert, and D. J. Simons. 2015. Inattentional blindness and individual differences in cognitive abilities. *PLOS ONE* 10 (8): e0134675.

Kreitz, C., P. Furley, D. Memmert, and D. J. Simons. 2016. The influence of attention set, working memory capacity, and expectations on inattentional blindness. *Perception* 45 (4): 386–399.

Lacey, D., and P. M. Salmon. 2015. It's dark in there: Using systems analysis to investigate trust and engagement in dark web forums. In *Proceedings of the 12th International Conference of Engineering Psychology and Cognitive Ergonomics Held as Part of HCI International 2015*, edited by D. Harris, pp. 117–128. Cham, Switzerland: Springer International Publishing.

Lane, R., N. A. Stanton, and D. Harrison. 2006. Applying hierarchical task analysis to medication administration errors. *Applied Ergonomics* 37 (5): 669–679.

Larsson, P., S. W. A. Dekker, and C. Tingvall. 2010. The need for a systems theory approach to road safety. *Safety Science* 48 (9): 1167–1174.

Larue, G. S., A. Rakotonirainy, and N. L. Haworth. 2016. A simulator evaluation of effects of assistive technologies on driver cognitive load at railway-level crossings. *Journal of Transportation Safety & Security* 8 (suppl 1): 56–69.

Lenné, M. G., V. Beanland, P. M. Salmon, A. J. Filtness, and N. A. Stanton. 2013. Checking for trains: An on-road study of what drivers actually do at level crossings. In *Rail Human Factors: Supporting Reliability, Safety and Cost Reduction*, edited by N. Dadashi, A. Scott, J. R. Wilson, and A. Mills, pp. 53–59. London: Taylor & Francis.

Lenné, M. G., C. M. Rudin-Brown, J. Navarro, J. Edquist, M. Trotter, and N. Tomasevic. 2011. Driver behaviour at rail level crossings: Responses to flashing lights, traffic signals and stop signs in simulated rural driving. *Applied Ergonomics* 42 (4): 548–554.

Leveson, N. 2004. A new accident model for engineering safer systems. *Safety Science* 42: 237–270.

Liedtka, J., and T. Ogilvie. 2010. *Designing for Growth: A Design Thinking Tool Kit for Managers*. New York: Columbia Business School Publishing.

Lim, K. Y., and J. Long. 1994. *The MUSE Method for Usability Engineering*. Cambridge: Cambridge University Press.

Lintern, G. 2005. Integration of cognitive requirements into system design. *Proceedings of the Human Factors and Ergonomics Society 49th Annual Meeting* 49: 239–243.

Liu, J., B. Bartnik, S. H. Richards, and A. J. Khattak. 2016. Driver behavior at highway–rail grade crossings with passive traffic controls: A driving simulator study. *Journal of Transportation Safety & Security* 8 (suppl 1): 37–55.

Lockton, D., D. Harrison, and N. A. Stanton. 2010. The design with intent method: A design tool for influencing user behaviour. *Applied Ergonomics* 41 (3): 382–392.

Macal, C. M., and M. J. North. 2010. Tutorial on agent-based modelling and simulation. *Journal of Simulation* 4: 151–162.

Madsen, K. H. 1994. A guide to metaphorical design. *The Communications of the ACM* 37 (12): 57–62.

Mayhew, D. R., H. M. Simpson, A. F. Williams, and S. A. Ferguson. 1998. Effectiveness and role of driver education and training in a graduated licensing system. *Journal of Public Health Policy* 19 (1): 51–67.

McClure, R. J., C. Adriazola-Steil, C. Mulvihill, M. Fitzharris, P. Salmon, P. Bonnington, and M. Stevenson. 2015. Simulating the dynamic effect of land use and transport policies on the health of populations. *American Journal of Public Health* 105 (S2): S223–S229.

McIlroy, R. C., and N. A. Stanton. 2015. Ecological interface design two decades on: Whatever happened to the SRK taxonomy? *IEEE Transactions on Human-Machine Systems* 45 (2): 45–163.

McKnight, A. J., and A. S. McKnight. 2003. Young novice drivers: Careless or clueless? *Accident Analysis & Prevention* 35 (6): 921–925.

McLean, J., P. Croft, N. Elazar, and P. Roper. 2010. *Safe Intersection Approach Treatments and Safer Speeds through Intersections: Final Report, Phase 1*. Sydney, Australia: Austroads.

Mehnert, A., I. Nanninga, M. Fauth, and I. Schäfer. 2012. Course and predictors of posttraumatic stress among male train drivers after the experience of 'person under the train' incidents. *Journal of Psychosomatic Research* 73 (3): 191–196.

Mendoza, P. A., A. Angelelli, and A. Lindgren. 2011. Ecological interface design inspired human machine interface for advanced driver assistance systems. *IET Intelligent Transport Systems* 5 (1): 53–59.

Militello, L. G., and R. J. B. Hutton. 1998. Applied cognitive task analysis (ACTA): A practitioner's toolkit for understanding cognitive task demands. *Ergonomics* 41 (11): 1618–1641.

Mitsopoulos, E., M. A. Regan, T. J. Triggs, E. Wigglesworth, and N. Tomasevic. 2002. Do pictogram signs increase safety at passive crossings? A simulator experiment. *Paper presented at the 7th International Symposium on Railroad-Highway Grade Crossing Research and Safety*, Melbourne, Victoria, Australia, February 20–21.

Morgan, G. 1980. Paradigms, metaphors, and puzzle solving in organization theory. *Administrative Science Quarterly* 25 (4): 605–622.

Most, S. B., D. J. Simons, B. J. Scholl, R. Jimenez, E. Clifford, and C. F. Chabris. 2001. How not to be seen: The contribution of similarity and selective ignoring to sustained inattentional blindness. *Psychological Science* 12 (1): 9–17.

Mulvihill, C. M., P. M. Salmon, V. Beanland, M. G. Lenné, N. A. Stanton, and G. J. M. Read. 2016. Using the decision ladder to understand road user decision making at actively controlled rail level crossings. *Applied Ergonomics* 56: 1–10.

Mumford, E. 2006. The story of socio-technical design: Reflections on its successes, failures and potential. *Information Systems Journal* 16 (4): 317–342.

Naikar, N. 2013. *Work Domain Analysis: Concepts, Guidelines and Cases*. Boca Raton, FL: Taylor & Francis.

Naikar, N., A. Moylan, and B. Pearce. 2006. Analysing activity in complex systems with cognitive work analysis: Concepts, guidelines and case study for control task analysis. *Theoretical Issues in Ergonomics Science* 7 (4): 371–394.

Naikar, N., B. Pearce, D. Drumm, and P. M. Sanderson. 2003. Designing teams for first-of-a-kind, complex systems using the initial phases of cognitive work analysis: Case study. *Human Factors* 45 (2): 202–217.

National Safety Council. 2016. *National Safety Council Motor Vehicle Fatality Estimates*. Itasca, IL: National Safety Council. Retrieved from http://www.nsc.org/NewsDocuments/2016/mv-fatality-report-1215.pdf (Accessed March 11, 2017).

National Health and Medical Research Council. 2007/2015. *National Statement on Ethical Conduct in Human Research*. Canberra, Australia: The National Health and Medical Research Council. Retrieved from https://www.nhmrc.gov.au/_files_nhmrc/publications/attachments/e72_national_statement_may_2015_150514_a.pdf (Accessed March 11, 2017).

Naweed, A. 2013. Psychological factors for driver distraction and inattention in the Australian and New Zealand rail industry. *Accident Analysis & Prevention* 60: 193–204.

Neisser, U. 1976. *Cognition and Reality*. San Francisco, CA: W.H. Freeman and Company.

Neville, T. J., P. M. Salmon, and G. J. M. Read. 2016. Analysis of in-game communication as an indicator of recognition primed decision making in elite Australian rules football umpires. *Journal of Cognitive Engineering and Decision Making* 1–6.

Newnam, S., and N. Goode. 2015. Do not blame the driver: A systems analysis of the causes of road freight crashes. *Accident Analysis & Prevention* 76: 141–151.

Norman, D. A. 2007. *The Design of Future Things*. New York: Basic Books.

Norros, L. 2014. Developing human factors/ergonomics as a design discipline. *Applied Ergonomics* 45 (1): 61–71.

OCI (Office of the Chief Investigator). 2007. *Level Crossing Collision V/Line Passenger Train 840 8042 and A Truck Near Kerang, Victoria, 5th June 2007. Rail Safety Investigation 841, Report No. 2007/09. 842*. Melbourne, Australia: Office of the Chief Investigator.

O'Hare, D., M. Wiggins, A. Williams, and W. Wong. 2000. Cognitive task analyses for decision centred design and training. In *Task Analysis*, edited by J. Annett and N. Stanton, pp. 170–190. London: Taylor & Francis.

ONRSR (Office of the National Rail Safety Regulator). 2015. *Rail Safety Report 2014–15*. Adelaide, South Australia: Office of the National Rail Safety Regulator.

Øvergård, K. I., L. J. Sorensen, S. Nazir, and T. J. Martinsen. 2015. Critical incidents during dynamic positioning: Operators' situation awareness and decision-making in maritime operations. *Theoretical Issues in Ergonomics Science* 16 (4): 366–387.

Pasmore, W., C. Francis, J. Haldeman, and A. Shani. 1982. Sociotechnical systems: A North American reflection on empirical studies. *Human Relations* 35 (12): 1179–1204.

Perrow, C. 1999. *Normal Accidents: Living with High-risk Technologies*. Princeton, NJ: Princeton University Press.

Phipps, D., P. Beatty, C. Nsoedo, and D. Parker. 2008. Human factors in anaesthetic practice: Insights from a task analysis. *British Journal of Anaesthesia* 100 (3): 333–343.

Plant, K. L., and N. A. Stanton. 2013. What is on your mind? Using the perceptual cycle model and critical decision method to understand the decision-making process in the cockpit. *Ergonomics* 56 (8): 1232–1250.

Potter, S. S., E. M. Roth, D. D. Woods, and W. C. Elm. 1998. A framework for integrating cognitive task analysis into the system development framework. *Proceedings of the Human Factors and Ergonomics Society Annual Meeting* 42: 395–399.

R v Scholl. 2009. Transcript of Proceedings. Supreme Court of Victoria, Kaye J, 855 (25 May) to 856 (13 June).

Rafferty, L., G. H. Walker, and N. A. Stanton. 2012. *The Human Factors of Fratricide.* Aldershot: Ashgate.

Rasmussen, J. 1986. *Information Processing and Human-Machine Interaction.* New York: North Holland.

Rasmussen, J., A. Pejtersen, and G. I. Goodstein. 1994. *Cognitive Systems Engineering.* New York: John Wiley & Sons.

Rasmussen, J. 1983. Skills, rules and knowledge – Signals, signs and symbols, and other distinctions in human performance models. *IEEE Transactions on Systems, Man and Cybernetics* 13: 257–266.

Rasmussen, J. 1997. Risk management in a dynamic society: A modelling problem. *Safety Science* 27: 183–213.

Rasmussen, J., A. M. Pejtersen, and K. Schmidt. 1990. *Taxonomy for Cognitive Work Analysis.* Roskilde, Denmark: Risø National Laboratory.

Read, G. J. M. 2015. Extension and application of cognitive work analysis to improve pedestrian safety at rail level crossings. PhD, Monash Injury Research Institute, Monash University, Clayton, Australia.

Read, G. J. M., P. M. Salmon, and M. G Lenné. 2012. From work analysis to work design: A review of cognitive work analysis design applications. *Proceedings of the Human Factors and Ergonomics Society Annual Meeting,* Boston, MA.

Read, G. J. M., P. M. Salmon, and M. G. Lenné. 2013. Sounding the warning bells: The need for a systems approach to understanding behaviour at rail level crossings. *Applied Ergonomics* 44 (5): 764–774.

Read, G. J. M., P. M. Salmon, and M. G. Lenné. 2015a. Cognitive work analysis and design: Current practice and future practitioner requirements. *Theoretical Issues in Ergonomics Science* 16: 154–173.

Read, G. J. M., P. M. Salmon, and M. G. Lenné. 2015b. The application of a systems thinking design toolkit to improve situation awareness and safety at road intersections. *Procedia Manufacturing* 3: 2613–2620.

Read, G. J. M., P. M. Salmon, and M. G. Lenné. 2016a. When paradigms collide at road and rail interface: Evaluation of a sociotechnical systems theory design toolkit for cognitive work analysis. *Ergonomics* 59 (9): 1135–1157.

Read, G. J. M., P. M. Salmon, M. G. Lenné, and E. M. Grey. 2014. Evaluating design hypotheses for rail level crossings: An observational study of pedestrian and cyclist behaviour. *Proceedings of the 5th International Conference on Applied Human Factors and Ergonomics,* Kraków, Poland.

Read, G. J. M., P. M. Salmon, M. G. Lenné, and D. P. Jenkins. 2015c. Designing a ticket to ride with the cognitive work analysis design toolkit. *Ergonomics* 58 (8): 1266–1286.

Read, G. J. M., P. M. Salmon, M. G. Lenné, and N. A. Stanton. 2015d. Designing sociotechnical systems with cognitive work analysis: Putting theory back into practice. *Ergonomics* 58 (5): 822–851.

Read, G. J. M., P. M. Salmon, M. G. Lenné, N. A. Stanton, C. M. Mulvihill, and K. L. Young. 2016b. Applying the prompt questions from the cognitive work analysis design toolkit: A demonstration in rail level crossing design. *Theoretical Issues in Ergonomics Science* 17 (4): 354–375.

Reason, J. 1997. *Managing the Risks of Organizational Accidents*. Aldershot: Ashgate.

Reason, J., A. Manstead, S. Stradling, J. Baxter, and K. Campbell. 1990. Errors and violations on the roads: A real distinction? *Ergonomics* 33 (10–11): 1315–1332.

Regan, M. A., K. Young, T. Triggs, N. Tomasevic, E. Mitsopoulos, P. Tierney, D. Healy, K. Connelly, and C. Tingvall. 2005. Effects on driving performance of in-vehicle intelligent transport systems: Final results of the Australian TAC SafeCar project. *Journal of the Australasian College of Road Safety* 18 (1): 23–30.

RISSB (Rail Industry Safety and Standards Board). 2009. *Level Crossing Stocktake*. Canberra, Australia: Rail Industry Safety and Standards Board. Retrieved from https://www.rissb.com.au/wp-content/uploads/2014/08/7_-National-Level-Crossing-Stocktake.pdf (Accessed March 11, 2017).

Road Safety Committee. 2008. *Report of the Road Safety Committee on the Inquiry into Improving Safety at Level Crossings*. Melbourne, Australia: Victorian Government Printer.

Robinson, G. H. 1982. Accidents and sociotechnical systems: Principles for design. *Accident Analysis and Prevention* 14 (2): 121–130.

Rudin-Brown, C. M., M. G. Lenné, J. Edquist, and J. Navarro. 2012. Effectiveness of traffic light vs. boom barrier controls at road–rail level crossings: A simulator study. *Accident Analysis & Prevention* 45: 187–194.

Saccomanno, F. F., P. Y.-J. Park, and L. Fu. 2007. Estimating countermeasure effects for reducing collisions at highway–railway grade crossings. *Accident Analysis & Prevention* 39 (2): 406–416.

Safe Work Australia. 2013. *The Incidence of Accepted Workers' Compensation Claims for Mental Stress in Australia*. Canberra, Australia: Safe Work Australia. Retrieved from http://www.safeworkaustralia.gov.au/sites/SWA/about/Publications/Documents/769/The-Incidence-Accepted-WC-Claims-Mental-Stress-Australia.pdf (Accessed March 11, 2017).

Salmon, P. M., A. Clacy, and C. Dallat. 2017. It's not all about the bike: Distributed situation awareness and teamwork in elite women's cycling teams. In *Contemporary Ergonomics and Human Factors 2017*. Boca Raton, FL: CRC Press.

Salmon, P. M., M. Cornelissen, and M. J. Trotter. 2012a. Systems-based accident analysis methods: A comparison of Accimap, HFACS, and STAMP. *Safety Science* 50 (4): 1158–1170.

Salmon, P. M., N. Goode, A. Spiertz, M. Thomas, E. Grant, and A. Clacy. 2016a. Is it really good to talk? Testing the impact of providing concurrent verbal protocols on driving performance. *Ergonomics* 12: 1–10.

Salmon, P. M., N. Goode, N. Taylor, M. G. Lenné, C. E. Dallat, and C. F. Finch. 2017. Rasmussen's legacy in the great outdoors: A new incident reporting and learning system for led outdoor activities. *Applied Ergonomics* 59, Part B: 637–648.

Salmon, P. M., and M. G. Lenné. 2009. Putting the 'system' into safe system frameworks. *Australasian College of Road Safety Journal* 30 (3): 21–22.

Salmon, P. M., and M. G. Lenné. 2015. Miles away or just around the corner: Systems thinking in road safety research and practice. *Accident Analysis and Prevention* 74: 243–249.

Salmon, P. M., M. G. Lenné, G. J. M. Read, C. M. Mulvihill, M. Cornelissen, G. H. Walker, K. L. Young, N. Stevens, and N. A. Stanton. 2016b. More than meets the eye: Using cognitive work analysis to identify design requirements for future rail level crossing systems. *Applied Ergonomics* 53: 312–322.

Salmon, P. M., M. G. Lenné, G. H. Walker, N. A. Stanton, and A. Filtness. 2014a. Exploring schema-driven differences in situation awareness between road users: An on-road study of driver, cyclist and motorcyclist situation awareness. *Ergonomics* 57 (2): 191–209.

Salmon, P. M., M. G. Lenné, G. H. Walker, N. A. Stanton, and A. Filtness. 2014b. Using the Event Analysis of Systemic Teamwork (EAST) to explore conflicts between different road user groups when making right hand turns at urban intersections. *Ergonomics* 57 (11): 1628–1642.

Salmon, P. M., M. G. Lenné, K. L. Young, and G. H. Walker. 2013a. A network analysis-based comparison of novice and experienced driver situation awareness at rail level crossings. *Accident Analysis and Prevention* 58: 195–205.

Salmon, P. M., R. McClure, and N. A. Stanton. 2012b. Road transport in drift? Applying contemporary systems thinking to road safety. *Safety Science* 50 (9): 1829–1838.

Salmon, P. M., G. J. M. Read, M. G. Lenné, and N. A. Stanton. 2013b. The shared responsibility for road safety: What is the road transport 'system' and who is the responsibility shared amongst? *Paper Presented at the Systems Engineering and Test and Evaluation Conference (SETE2013)*, Canberra, Australia, April 29–May 1.

Salmon, P. M., G. J. M. Read, N. A. Stanton, and M. G Lenné. 2013c. The Crash at Kerang: Investigating systemic and psychological factors leading to unintentional non-compliance at rail level crossings. *Accident Analysis and Prevention* 50: 1278–1288.

Salmon, P. M., N. A. Stanton, A. C. Gibbon, D. P. Jenkins, and G. H. Walker. 2010. *Human Factors Methods and Sports Science: A Practical Guide.* Boca Raton, FL: Taylor & Francis.

Salmon, P. M., K. L. Young, and M. Cornelissen. 2013d. Compatible cognition amongst road users: The compatibility of driver, motorcyclist, and cyclist situation awareness. *Safety Science* 56: 6–17.

Salmon, P. M., K. L. Young, and M. A. Regan. 2011. Distraction 'on the buses': A novel framework of ergonomics methods for identifying sources and effects of bus driver distraction. *Applied Ergonomics* 42 (4): 602–610.

Sanders, E. B. N. 2002. From user-centered to participatory design approaches. In *Design and the Social Sciences: Making Connections*, edited by J. Frascara, pp. 1–8. London: Taylor & Francis.

Sandin, J. 2009. An analysis of common patterns in aggregated causation charts from intersection crashes. *Accident Analysis & Prevention* 41 (3): 624–632.

Sarvi, M., and M. Kuwahara. 2007. Microsimulation of freeway ramp merging processes under congested traffic conditions. *IEEE Transactions on Intelligent Transport Systems* 8 (3): 470–479.

Scott-Parker, B., N. Goode, and P. M. Salmon. 2015. The driver, the road, the rules … and the rest? A systems approach to young driver safety. *Accident Analysis & Prevention* 74: 297–305.

Shepherd, A. 1989. Analysis and training in information technology tasks. In *Task Analysis for Human-Computer Interaction*, edited by D. Diaper, pp. 15–55. Chichester: Ellis Horwood.

Shepherd, A. 2002. *Hierarchical Task Analysis.* London: Taylor & Francis.

Shepherd, S. P. 2014. A review of system dynamics models applied in transportation. *Transportmetrica B: Transport Dynamics* 2 (2): 83–105.

Sinclair, M. A. 2007. Ergonomics issues in future systems. *Ergonomics* 50 (12): 1957–1986.

Snowden, D. J., and M. E. Boone. 2007. A leader's framework for decision making. *Harvard Business Review* 85: 68–76.

Standards Australia. 2007. *Manual of Uniform Traffic Control Devices. Part 7: Railway Crossings* (AS 1742.7-2007). Sydney, Australia: Standards Australia.

Stanton, N. A. 2006. Hierarchical task analysis: Developments, applications, and extensions. *Applied Ergonomics* 37 (1): 55–79.

Stanton, N. A. 2014. Representing distributed cognition in complex systems: How a submarine returns to periscope depth. *Ergonomics* 57 (3): 403–418.

Stanton, N. A., and K. Bessell. 2014. How a submarine returns to periscope depth: Analysing complex socio-technical systems using Cognitive Work Analysis. *Applied Ergonomics* 45 (1): 110–125.

Stanton, N. A., D. Harris, and A. Starr. 2016. The future flight deck: Modelling dual, single and distributed crewing options. *Applied Ergonomics* 53, Part B: 331–342.

Stanton, N. A., R. C. McIlroy, C. Harvey, S. Blainey, A. Hickford, J. M. Preston, and B. Ryan. 2013a. Following the cognitive work analysis train of thought: Exploring the constraints of modal shift to rail transport. *Ergonomics* 56 (3): 522–540.

Stanton, N. A., P. Salmon, D. Harris, A. Marshall, J. Demagalski, M. S. Young, T. Waldmann, and S. Dekker. 2009. Predicting pilot error: Testing a new methodology and a multi-methods and analysts approach. *Applied Ergonomics* 40 (3): 464–471.

Stanton, N. A., P. M. Salmon, L. A. Rafferty, G. H. Walker, C. Baber, and D. P. Jenkins. 2013b. *Human Factors Methods: A Practical Guide for Engineering and Design*. 2nd ed. Aldershot: Ashgate.

Stanton, N. A., P. M. Salmon, G. H. Walker, and D. P. Jenkins. 2017. *Cognitive Work Analysis: Applications, Extensions and Future Directions*. Boca Raton, FL: CRC Press.

Sterman, J. D. 2000. *Business Dynamics: Systems Thinking and Modeling for a Complex World*. New York: Irwin.

Svedung, I., and J. Rasmussen. 2002. Graphic representation of accident scenarios: Mapping system structure and the causation of accidents. *Safety Science* 40 (5): 397–417.

Svedung, I., and J. Rasmussen. 2000. *Proactive Risk Management in a Dynamic Society*. Karlstad, Sweden: Swedish Rescue Services Agency.

Tey, L. S., L. Ferreira, and A. Wallace. 2011. Measuring driver responses at railway level crossings. *Accident Analysis & Prevention* 43 (6): 2134–2141.

Tey, L. S., G. Wallis, S. Cloete, L. Ferreira, and S. Zhu. 2013. Evaluating driver behavior toward innovative warning devices at railway level crossings using a driving simulator. *Journal of Transportation Safety & Security* 5 (2): 118–130. doi:10.1080/19439962.2012.731028.

Thompson, J., S. Newnam, and M. Stevenson. 2015. A model for exploring the relationship between payment structures, fatigue, crash risk, and regulatory response in a heavy-vehicle transport system. *Transportation Research Part A* 82: 204–215.

Tichon, J. 2007. The use of expert knowledge in the development of simulations for train driver training. *Cognition, Technology & Work* 9 (4): 177–187.

Transportation Safety Board of Canada. 1996. *Trespasser Fatality, VIA Rail Canada Inc. Train No. 76, Mile 98.65, Chatham Subdivision, Tecumseh, Ontario*. Quebec, Canada: Transportation Safety Board of Canada.

Trist, E. L., and K. W. Bamforth. 1951. Some social and psychological consequences of the longwall method of coal-getting: An examination of the psychological situation and defences of a work group in relation to the social structure and technological content of the work system. *Human Relations* 4 (1): 3–38.

Underwood, G. 2007. Visual attention and the transition from novice to advanced driver. *Ergonomics* 50 (8): 1235–1249.

Underwood, G., P. Chapman, K. Bowden, and D. Crundall. 2002. Visual search while driving: Skill and awareness during inspection of the scene. *Transportation Research Part F: Traffic Psychology and Behaviour* 5 (2): 87–97.

Vicente, K. J. 1999. *Cognitive Work Analysis: Toward Safe, Productive, and Healthy Computer-Based Work*. Mahwah, NJ: Lawrence Erlbaum Associates.

Vicente, K. J. 2002. Ecological interface design: Process and challenges. *Human Factors* 44 (1): 62–78.

Vicente, K., and K. Christoffersen. 2006. The Walkerton E. Coli outbreak: A test of Rasmussen's framework for risk management in a dynamic society. *Theoretical Issues in Ergonomics Science* 7 (2): 93–112.

Vicente, K. J., and J. Rasmussen. 1990. The ecology of human-machine systems II: Mediating 'direct perception' in complex work domains. *Ecological Psychology* 2 (3): 207–249.

Vicente, K. J., and J. Rasmussen. 1992. Ecological interface design: Theoretical foundations. *IEEE Transactions on Systems, Man & Cybernetics* 22 (4): 589–606.

VicRoads. 2013. CrashStats. Retrieved from http://www.vicroads.vic.gov.au/Home/SafetyAndRules/AboutRoadSafety/StatisticsAndResearch/CrashStats.htm (Accessed December 16, 2013).

Walker, G. H. 2005. Verbal protocol analysis. In *Handbook of Human Factors methods*, edited by N. A. Stanton, A. Hedge, K, Brookhuis, E. Salas, and H. Hendrick, pp. 30.1–30.9. Boca Raton, FL: CRC Press.

Walker, G. H., N. A. Stanton, T. A. Kazi, P. M. Salmon, and D. P. Jenkins. 2009a. Does advanced driver training improve situational awareness? *Applied Ergonomics* 40 (4): 678–687.

Walker, G. H., N. A. Stanton, and P. M. Salmon. 2011. Cognitive compatibility of motorcyclists and car drivers. *Accident Analysis and Prevention* 43: 878–888.

Walker, G. H., N. A. Stanton, and P. M. Salmon. 2015. *Human Factors in Automotive Engineering and Technology*. Aldershot: Ashgate.

Walker, G. H., N. A. Stanton, P. M. Salmon, and D. P. Jenkins. 2009b. *Command and Control: The Sociotechnical Perspective*. Aldershot: Ashgate.

Watson, M. O., and P. M. Sanderson. 2007. Designing for attention with sound: Challenges and extensions to ecological interface design. *Human Factors* 49: 331–346.

Weber, R. P. 1990. *Basic Content Analysis*. London: Sage Publications.

Wigglesworth, E., and C. B. Uber. 1991. An evaluation of the railway level crossing boom barrier program in Victoria, Australia. *Journal of Safety Research* 22 (3): 133–140.

Wong, B. L. W. 2004. Critical decision method data analysis. In *The Handbook of Task Analysis for Human-Computer Interaction*, edited by D. Diaper and N. A. Stanton, pp. 327–346. Mahwah, NJ: Lawrence Erlbaum.

World Health Organisation. 2014. *The Top Ten Causes of Death*. Retrieved from http://www.who.int/mediacentre/factsheets/fs310/en/ (Accessed March 11, 2017).

Wright, T. J., W. R. Boot, and C. S. Morgan. 2013. Pupillary response predicts multiple object tracking load, error rate, and conscientiousness, but not inattentional blindness. *Acta Psychologica* 144 (1): 6–11.

Wullems, C. 2011. Towards the adoption of low-cost rail level crossing warning devices in regional areas of Australia: A review of current technologies and reliability issues. *Safety Science* 49 (8): 1059–1073.

Wullems, C., P. Hughes, and G. Nikandros. 2013. Modelling risk at low exposure railway level crossings: Supporting an argument for low-cost level crossing warning devices with lower levels of safety integrity. *Proceedings of the Institution of Mechanical Engineers, Part F: Journal of Rail and Rapid Transit* 227 (5): 560–569.

Young, K. L., M. G. Lenné, V. Beanland, P. M. Salmon, and N. A. Stanton. 2015. Where do novice and experienced drivers direct their attention on approach to urban rail level crossings? *Accident Analysis and Prevention* 77: 1–11.

Young, K. L., and P. M. Salmon. 2015. Sharing the responsibility for driver distraction across road transport systems: A systems approach to the management of distracted driving. *Accident Analysis and Prevention* 74: 350–359.

Young, K. L., P. M. Salmon, and M. Cornelissen. 2013a. Distraction-induced driving error: An on-road examination of the errors made by distracted and undistracted drivers. *Accident Analysis and Prevention* 58: 218–225.

Young, K. L., P. M. Salmon, and M. Cornelissen. 2013b. Missing links? The effects of distraction on driver situation awareness. *Safety Science* 56: 36–43.

Young, K. L., P. M. Salmon, and M. G. Lenné. 2013. At the cross-roads: An on-road examination of driving errors at intersections. *Accident Analysis and Prevention* 58: 226–234.

Young, M. S., and S. A. Birrell. 2012. Ecological IVIS design: Using EID to develop a novel in-vehicle information system. *Theoretical Issues in Ergonomics Science* 13 (2): 225–239.

Index

Note: Page numbers followed by f and t refer to figures and tables, respectively.